iPhone®
Application Development
FOR
7-11

DEMCO

iPhone® Application Development For Dummies,® 3rd Edition

Published by
Wiley Publishing, Inc.
111 River Street
Hoboken, NJ 07030-5774

www.wiley.com

Copyright © 2010 by Wiley Publishing, Inc., Indianapolis, Indiana

Published by Wiley Publishing, Inc., Indianapolis, Indiana

Published simultaneously in Canada

For general information on our other products and services, please contact our Customer Care Department within the U.S. at 877-762-2974, outside the U.S. at 317-572-3993, or fax 317-572-4002.

For technical support, please visit www.wiley.com/techsupport.

Wiley also publishes its books in a variety of electronic formats. Some content that appears in print may not be available in electronic books.

Library of Congress Control Number: 2010935563

ISBN: 978-0-470-87996-2

Manufactured in the United States of America

10 9 8 7 6 5 4 3 2 1

WILEY

About the Author

Neal Goldstein is a recognized leader in making state-of-the-art and cutting-edge technologies practical for commercial and enterprise development. He was one of the first technologists to work with commercial developers at firms such as Apple Computer, Lucasfilm, and Microsoft to develop commercial applications using object-based programming technologies. He was a pioneer in moving that approach into the corporate world for developers at Liberty Mutual Insurance, USWest (now Verizon), National Car Rental, EDS, and Continental Airlines, showing them how object-oriented programming could solve enterprise-wide problems. His book (with Jeff Alger) on object-oriented development, *Developing Object-Oriented Software for the Macintosh* (Addison Wesley, 1992), introduced the idea of scenarios and patterns to developers. He was an early advocate of the Microsoft .NET framework, and he successfully introduced it into many enterprises, including Charles Schwab. He was one of the earliest developers of Service Oriented Architecture (SOA), and as Senior Vice President of Advanced Technology and the Chief Architect at Charles Schwab, he built an integrated SOA solution that spanned the enterprise, from desktop PCs to servers to complex network mainframes. (He holds three patents as a result.) As one of IBM's largest customers, he introduced the folks at IBM to SOA at the enterprise level and encouraged them to head in that direction.

He is passionate about the real value mobile devices can provide and has eight applications in the App Store, These include a series of Travel Photo Guides (`http://travelphotoguides.com`) developed with his partners at mobilefortytwo and a Digital Field Guides series (`http://lp.wileypub.com/DestinationDFGiPhoneApp`) developed in partnership with John Wiley & Sons. He also has a cool little free app called Expense Calendar that allows you to keep track of things like expenses, mileage, and time by adding them to your calendar.

Along with those apps, he has written several books on iPhone programming, including *iPhone Application Development For Dummies* (Wiley), *iPhone Application Development For Dummies* (2nd Edition) (Wiley), and *Objective-C For Dummies* (Wiley), and he co-authored (with Tony Bove) *iPhone Application Development All-in-One For Dummies* (Wiley) and *iPad Application Development For Dummies* (Wiley). He's also the co-author (with Jon Manning and Paris Buttfield-Addison) of the forthcoming *iPhone & iPad Game Development For Dummies*.

Because you can never tell what he'll be up to next, check regularly at his Web site: `www.nealgoldstein.com`.

Dedication

To my children Sarah and Evan, and all of my personal and artist friends who have kept me centered on the (real) world outside of writing and technology. But most of all to my wife, Linda, who is everything that I ever hoped for and more than I deserve. Yes, Sam . . . the light at the end of the tunnel is not a freight train.

Author's Acknowledgments

There is no better Acquisitions Editor than Katie Feltman, who does a superb job of keeping me on track and doing whatever she needed to do to allow me to stay focused on the writing. Paul Levesque is the Project Editor's Project Editor, and has been known to do even more than six impossible things before breakfast. Copy Editor Virginia Sanders has done yet another great job in helping me make things clearer, and has a great sense of humor. Thanks again to my agent Carole Jelen, for her continued work and support in putting these projects together.

Publisher's Acknowledgments

We're proud of this book; please send us your comments at http://dummies.custhelp.com. For other comments, please contact our Customer Care Department within the U.S. at 877-762-2974, outside the U.S. at 317-572-3993, or fax 317-572-4002.

Some of the people who helped bring this book to market include the following:

Acquisitions, Editorial, and Media Development

Senior Project Editor: Paul Levesque

Acquisitions Editor: Katie Feltman

Copy Editor: Virginia Sanders

Technical Editor: Glenda Adams

Editorial Manager: Leah Cameron

Media Development Project Manager: Laura Moss-Hollister

Media Development Assistant Project Manager: Jenny Swisher

Media Development Associate Producers: Josh Frank, Marilyn Hummel, Douglas Kuhn, and Shawn Patrick

Editorial Assistant: Amanda Graham

Sr. Editorial Assistant: Cherie Case

Cartoons: Rich Tennant (www.the5thwave.com)

Composition Services

Project Coordinator: Patrick Redmond

Layout and Graphics: Timothy C. Detrick, Christine Williams

Proofreaders: Rebecca Denoncour, Bonnie Mikkelson

Indexer: Potomac Indexing, LLC

Publishing and Editorial for Technology Dummies

　　Richard Swadley, Vice President and Executive Group Publisher

　　Andy Cummings, Vice President and Publisher

　　Mary Bednarek, Executive Acquisitions Director

　　Mary C. Corder, Editorial Director

Publishing for Consumer Dummies

　　Diane Graves Steele, Vice President and Publisher

Composition Services

　　Debbie Stailey, Director of Composition Services

Contents at a Glance

Table of Contents

Part IV: An Industrial-Strength Application 267

Introduction

A lot has changed since I put pen to paper (okay, finger to keyboard) and started writing the first edition of this book back in 2008. The newest iPhone — with its Retina display, 5.0 megapixel front-facing still camera with LED flash, HD video recording, and gyroscope — is truly an amazing piece of hardware, light years ahead of the original iPhone, which (admittedly) was pretty cool to begin with. iOS 4 is a game-changing advancement over iPhone OS 2, becoming both *broader* and *deeper* — broader in the amount of functionality offered and deeper in the control you have over that functionality. And then of course there is multitasking.

But one thing hasn't changed. In the first edition of *iPhone Application Development For Dummies,* I said that when Apple opened up the iPhone to developers, I got as excited about developing software as I did when I first discovered the power of the Mac. And you know what? I'm still excited.

As I continue to explore the iPhone as a new platform, I keep finding more possibilities for applications that never existed before. The iPhone is a mobile computer, but it's not simply a mobile desktop. Its hardware and software make it possible to wander the world, or your own neighborhood, and stay connected to whomever and whatever you want to. It enables a new class of here-and-now applications that allow you to do what you need to, based on what's going on around you and where you are.

The first edition of *iPhone Application Development For Dummies* was based on iPhone OS 2.2.1. When iPhone OS 3.0 was released, and then quickly followed by OS 3.1, I knew that I had to do a second edition. You're now reading the third edition, based on the new iPhone OS — iOS 4. With the new OS came a bunch of important new features that I wanted to show readers how to use. Multitasking, to take just one example, allows applications to continue to process events in the background and deliver local notifications to the user. (You'll definitely hear more about multitasking as you make your way through the book.)

The new OS also brought with it changes to the nuts and bolts of how to develop iPhone applications that required rewriting several chapters as well as rethinking the examples I roll out in order to show how to use the SDK to create real applications. The new gesture recognizers allow you, as an application developer, to process touches in the same way the iOS and built-in applications do. The Location Services framework now provides a way to reduce the power used to keep the user informed to where he or she is, and along with some added MapKit functionality, it's possible for even a beginning developer to take full advantage of the location hardware. (I rewrote one chapter and added a new chapter to show you exactly how.) To round things off, Apple came up with some changes to Xcode, its nifty suite of software development tools, and more important, its provisioning process. (Changes to the latter have made it much easier to get your applications up and running on the iPhone.)

As a result, this new edition is based on iOS 4 *and* Xcode 3.2.3. If you want to find out how to develop applications, this is the set of tools you absolutely need to use to do it the right way.

All the new features (and extensions of old features) bundled into iOS 4 are great and exciting and visionary, but all that greatness/excitement/visionary-ness comes at a cost. Way back in 2008, when I first started writing about iPhone application development, it was really difficult to get my head around the whole thing, conceptually speaking; sometimes I found it difficult to figure out exactly how to adapt my vision of how I thought my application should work with the way Apple thought my application should work. iOS 4 brought a whole new set of expectations into the mix.

Back in 2008, there were lots of resources out there quite willing to walk me through the Apple mindset, but to be honest that was precisely the problem: There were *lots* of resources! As in, *thousands* of pages of documentation I could read, and lots of sample code to look at. I could only get through a small fraction of the documentation before I just couldn't stand the suspense anymore and started coding. Naturally enough, there were a few false starts and blind alleys until I found my way, and it has been (pretty much) smooth sailing ever since.

That's why, when the *For Dummies* folks first asked me to write a book on developing software for the iPhone, I jumped at the chance. Here was an opportunity for me to write the book I wish I'd had when I started developing iPhone software.

But now it's even worse. As I said, iOS 4 is both broader and deeper and there are many more pages of documentation and sample code. That's why, when the *For Dummies* folks came around again and asked me to write a third edition of *iPhone Software Development For Dummies,* I dusted off my typewriter (just kidding) and got to work.

About This Book

iPhone Application Development For Dummies is a beginner's guide to developing iPhone applications. And not only do you *not* need any iPhone development experience to get started, you don't need any Macintosh development experience either. I expect you to come as a blank slate, ready to be filled with useful information and new ways to do things.

Because of the nature of the iPhone, you can create small, bite-sized applications that can be really powerful. And because you can start small and create real applications that do something important for a user, it's relatively easy to transform yourself from "I know nothing" into a developer who, though not (yet) a superstar, can still crank out quite a respectable application.

But the iPhone can be home to some pretty fancy software as well — so I take you on a journey through building an industrial-strength application and show you the ropes for developing one on your own.

This book distills the hundreds (or even thousands) of pages of Apple documentation, not to mention my own development experience, into only what's necessary to start you developing real applications. But this is no recipe book that leaves it up to you to put it all together; rather, it takes you through the frameworks (the code supplied in the SDK) and iPhone architecture in a way that gives you a solid foundation in how applications really work on the iPhone — and acts as a road map to expand your knowledge as you need to.

I assume that you're in this for the long haul and you want to master the whole application-development ball of wax. I use real-world applications to show the concepts and give you the background on how things really work on the iPhone — the in-depth knowledge you need to go beyond the simple "Hello World" apps and create those killer iPhone applications. So be prepared! There may be some places where you might want to say, "Get on with it," but — based on my experience (including eight apps in the App Store, seven books (and counting), and untold hours expended on in-person classes and technical talks — I'm giving you what you need to move from following recipes in a cookbook by rote to modifying and even creating your own recipes.

It's a multicourse banquet, intended to make you feel satisfied (and really full) at the end.

Conventions Used in This Book

This book guides you through the process of building iPhone applications. Throughout, you use the provided iPhone framework classes (and create new ones, of course) and code them by using the Objective-C programming language.

Code examples in this book appear in a monospaced font so they stand out a bit better. That means the code you see will look like this:

```
#import <UIKit/ UIKit.h>
```

Objective-C is based on C, which (I want to remind you) *is* case-sensitive, so please enter the code that appears in this book *exactly* as it appears in the text. I also use the standard Objective-C naming conventions — for example, class names always start with a capital letter, and the names of methods and instance variables always start with a lowercase letter.

Let me throw out that all URLs in this book appear in a monospaced font as well:

```
www.nealgoldstein.com
```

If you're ever uncertain about anything in the code, you can always look at the source code at the Web site associated with this book or on my Web site at `www.nealgoldstein.com` — from time to time, I provide updates for the code there, and post other things you might find useful. (You can grab the same material from the *For Dummies* Web site at `www.dummies.com/go/iphoneappdevfd3e`)

Note: As of this writing (in July 2010), Apple had released an Xcode 4 developer preview. When the final product is delivered, you'll find a brief overview on my Web site of how to use it with this book

Foolish Assumptions

To begin programming your iPhone applications, you need an Intel-based Macintosh computer with the latest version of the Mac OS on it. (No, you can't program iPhone applications on the iPhone.) You also need to download the iPhone Software Development Kit (SDK) — which is free — but you do have to become a registered iPhone developer before you can do that. (Don't worry; I show you how to do both.) And, oh yeah, you need an iPhone. You won't start running your application on it right away — you'll use the

Simulator that Apple provides with the iPhone SDK during the initial stages of development — but at some point, you'll want to test your application on a real, live iPhone.

I'm going to assume that you have some programming knowledge and that you have at least a passing acquaintance with object-oriented programming, using some variant of the C language (such as C++, C#, or maybe even Objective-C). In case you don't, I point out some resources that can help you get up to speed. The examples in this book are focused on the frameworks that come with the SDK; the code is pretty simple (usually) and straightforward. (I don't use this book as a platform to dazzle you with fancy coding techniques.)

I also assume that you're familiar with the iPhone itself and that you've at least explored Apple's included applications to get a good working sense of the iPhone look and feel. It would also help if you browse the App Store to see the kinds of applications available there and maybe even download a few free ones (as if I could stop you).

How This Book Is Organized

iPhone Application Development For Dummies, 3rd Edition, has five main parts.

Part I: Getting Started

Part I introduces you to the iPhone world. You find out what makes a great iPhone application, and how an iPhone application is structured. You also find out how to become an "official" iPhone developer and what you need to do to in order to be able to distribute your applications through Apple's App Store.

Part II: Using the iPhone Development Tools

I start Part II by showing you how to download the Software Development Kit (SDK) — and then help you unpack all the goodies contained therein, including Xcode (Apple's development environment for the OS X operating system) and Interface Builder. (You'll soon discover that the latter is more than your run-of-the-mill program for building graphical user interfaces.) I also explain how everything works together at runtime, which should give you a real feel for how an iPhone application works. Parts I and II give you the fundamental background that you need to develop iPhone applications.

Part III: From "Gee, That's a Good Idea," to the App Store

With the basics behind you and a good understanding of the application architecture under your belt, it's finally time to have some fun doing something useful. In this part, I show you how to create a simple application that people can actually use — if you lose your iPhone, it displays a phone number that some Good Samaritan can call to contact you. (My friends thought it was sheer genius.) What's more, the Good Samaritan who finds your phone only has to tap that number where it's shown onscreen, and the phone dials the number automatically. Putting this handy little app together will give you some practice at creating a useful, single-screen program with controls. It's a great application to introduce you to iPhone development — big enough to be useful, but small enough not to make your head explode.

It's also a "real" application. I go through the process I used to (successfully) submit it to the App Store, and you can download it and see it for yourself.

Part IV: An Industrial-Strength Application

Part IV takes you into the world of applications that contain major functionality. I show you how to design an application with lots of data, views, and access to the Web. I don't go slogging through every detail, but I demonstrate almost all the technology you need to master if you're going to create a compelling application like this on your own. I also spend a lot of time on Location Services and how to keep the user informed of where he or she is both on a map and through local notifications that are even generated when the app is in the background.

Part V: The Part of Tens

Part V consists of some tips to help you avoid having to discover everything the hard way. It talks about approaching application development in an "adult" way right from the beginning (without taking the fun out of it, I assure you). I also take you on a tour of the iPhone sample code, pointing out some samples I really like and have found to be the most useful.

Icons Used in This Book

This icon indicates a useful pointer that you shouldn't skip.

This icon represents a friendly reminder. It describes a vital point that you should keep in mind while proceeding through a particular section of the chapter.

This icon signifies that the accompanying explanation may be informative (dare I say, interesting?), but it isn't essential to understanding iPhone application development. Feel free to skip past these tidbits if you like (though skipping while leaning may be tricky).

This icon alerts you to potential problems that you may encounter along the way. Read and obey these blurbs to avoid trouble.

Where to Go from Here

It's time to explore the iPhone! If you're nervous, take heart: I've gotten lots of e-mails (and an occasional card and letter) from readers ranging in age from 12 to 67 that tell me how they're doing fine (and yes, I do respond to them all).

I also want to direct your attention to an app that I came up with that's now in the App Store. It's called Expense Calendar and I came up with the idea while writing this book. Download it (it's free) and check it out. What's interesting about it is that it took me three days to do, and 80 percent of the code I used is based on what you discover in this book. (The other 20 percent is based on code you can find in the second edition of the *iPhone Application Development For Dummies All-In-One* [Wiley].)

Go have some fun!

Part I
Getting Started

The 5th Wave   By Rich Tennant

"What I'm doing should clear your sinuses, take away your headache, and charge your iPhone."

In this part . . .

So you've decided you want to develop some software for the iPhone. You have a good idea for a utility — one that lets you know your net worth in Zimbabwean dollars, or one that acts as a data-driven application (say, one that knows where to find the best coffee in Seattle). Now what?

This part of the book lays out what you need to know to get started on the development journey. First of all, what makes a great iPhone application? Knowing that, you can evaluate your idea, see how it ranks, and maybe figure out what you have to do to transform it into something that knocks your users' socks off. Next, before you can actually build that sucker, you look under the hood at how iPhone applications work — what goes on behind the screen that ends up with a user seeing something in a window and interacting with controls. You get a look at the user interface frameworks and how to use them (and how they want to use you). Finally, to get all that free development software from Apple and to get your application into the App Store, you have to become "legal" — it's time to become an official iPhone developer.

Chapter 1

Creating Killer iPhone Applications

In This Chapter

▶ Figuring out what makes an insanely great iPhone application

▶ Listing the features of the iPhone that can inspire you

▶ Facing the limitations you have to live with

▶ Checking out the possibilities that are open to you

▶ Developing iPhone software now rather than later

*I*magine that you've just landed at Heathrow Airport. It's early in the morning, and you're dead tired as you clear customs. All you want to do now is find the fastest way to get into London, check into your hotel, and sleep for a few hours.

You take out your iPhone and touch the MobileTravel411 icon. On the left in Figure 1-1, you can see it asks whether you want to use Heathrow Airport as your current location. You touch Yes, and then touch Getting To From (as you can see in the center of Figure 1-1). Because it already knows that you're at Heathrow, it gives you your alternatives. Because of the congestion in and out of London, it suggests using the Heathrow Express, especially during rush hour.

You touch the Heathrow Express tab, and it tells you where to get the train and also tells you that the fare is £14.50 if you buy it from the ticket machine and £17.50 if you buy it onboard the train. (The iPhone on the right in Figure 1-1 is proof that I'm not making this up.) It turns out that you're so jetlagged that you can't do the math in your head, so you touch the Currency button, and it tells you that £ 14.50 is around $21.35 if you take it from the ATM, $21.14 on your no-exchange-rate-fee credit card, or $22.31 at the *bureau de change* at the airport.

Another touch gets you the current weather, which prompts you to dig out a sweater from your luggage before you get on the train.

When you get to Paddington Station, you really don't have a clue where the hotel that someone at the office booked for you might be. You touch Getting Around, and the application allows you to use the hotel address that is in your iPhone Contacts, and then gives you your options when it comes to finally finding that big, comfortable, English bed. These options include walking, taking a taxi, and public transit. You touch Tube, and it directs you to the nearest Tube stop and then displays fares, schedules, and how to buy a ticket.

Figure 1-1:
The Mobile
Travel411
application
can use
your current
location.

How much of a fantasy is this?

Not much. Most of this application already exists. What's more, it took me only a little more than three months to develop that application, starting from where you are now, with no iPhone programming experience.

Creating a Compelling User Experience

What users are after — and what devices like the iPhone enable — is when the separation between them and

- ✔ Other people
- ✔ Tools and information
- ✔ The application itself

is reduced to almost nothing.

These are the three dimensions of what I call "zero degrees of separation."

We (at least most of us) are social animals, and we feel a need to stay connected with friends, family, and yes, even business associates. What's truly amazing about the iPhone is that its feature set — its hardware and applications — enables this connectivity seamlessly, ranging from the phone itself to SMS to FaceTime (the iPhone's new video chat feature) to the built-in camera. Add to that the social networking applications available for your iPhone and your separation from other people reduces even further.

Although one could argue that our present ability to always connect could mean that we have too much of a good thing (especially when it comes to business associates), I'd say that the annoyance factor is a result of the lack of maturity in how we're using technology. If you think about it, the technology hasn't really been around all that long, and like most teenagers, it will mature over the next few years.

Many of the applications I've just mentioned have long been available on the desktop; what a device like the iPhone does is add mobility to the party. Not only can *you* connect from anywhere, but so can all the people you want to connect to.

This ability to run applications on your phone wherever you are makes it possible to have the information you need (as well as the tools you'd like to use) constantly available. But it's not just about the fact that the application you need is ready-to-run right there on your phone; it's (as important) about how the application is designed and implemented. It needs to "work right," requiring as little as possible from you in terms of effort when it comes to delivering to you what you need.

So, having the app is one thing, but having an optimally designed app is another. These first two dimensions are about what I call *content* — what an application actually does. Another aspect here, of course, is that the potential pool of stuff a user wants to be closer to depends on the *context* in which he's using the application — where she is and what is going on around a user. In that respect, I like to think of the here-and-nowness — the *relevance* — of the application and information. In plain terms, this means that you want to do a specific task with the help of up-to-date information, which the iPhone can easily access over the Internet through a cell network or Wi-Fi connection. You may even want the information or tasks tailored to where you are, which the iPhone can determine with its location hardware. To use a concrete example, a guidebook application may have a great user interface, for example, but it may not give me the most up-to-date information, or let me know a tour of Parliament is leaving in five minutes from the main entrance. Without those added touches, I'm just not willing to consider an app compelling.

The final dimension is the application itself, or more precisely, how the user interacts with the application. The Multi-Touch user experience sets it up so that the user is naturally more connected to the device — there is no mouse or keyboard acting as an intermediary — and what's more, Apple promotes the use of gestures as much as possible rather than the use of controls. If you want to move a map annotation, don't show arrow or direction controls or force the user type in a new address (although sometimes that may be the way to go) — let the user drag it to where she wants it to be, and provide feedback along the way.

The iPhone allows an immediacy and intimacy as it blends mobility and the power of the desktop to create a new kind of freedom. I like to use the term *user experience* because it implies more than a pretty user interface and nice graphics. A *compelling* user experience enables users to do what they need to do with a minimum of fuss and bother. But more than that, it forces you to think past a clean interface and even beyond basic convenience (such as not having to scroll through menus to do something simple).

Compelling Content in Context — What the App Does

There are a lot of different kinds of applications on the iPhone, ranging from Utility applications like Weather to communication-heavy ones like FaceTime to others that connect you to information and computing power on the Web or even to social networking communities.

These applications shine when they enable you to do what you'd never want to do on the desktop (or even a laptop), either because you don't have the hardware or because, even if you *did* have the hardware, doing it that way would be way too inconvenient. (The added bonus to such applications is that they're available no matter where you are.)

Imagine being at Heathrow, dead tired, taking out your laptop in the middle of a crowded terminal, powering it up, launching the application, and then navigating through it with the touchpad to get the information I got easily while holding the iPhone in one hand. I want that kind of information quickly and conveniently; I don't want to have to dig my way to it through menus or layers of screens (or even going through the hassle of finding a wireless Internet connection). Seconds count. By the time any road warrior tied to a laptop did this at Heathrow, I would already be on the Heathrow Express.

And don't forget games and other pieces of frivolous (that's the point) entertainment to while away the time while waiting in line.

I know what I want, and I want it now, and by the way that's all I want

There are three important things here I want to highlight: relevance, relevance, relevance.

What most of the really good iPhone applications have in common is focus. They address a well-defined task that can be done within a time span that is appropriate for that task. If I need to look something up, I want it right now! If I am playing a game while waiting in line, I want it to be of short duration, or broken up into a series of short and entertaining steps.

The application content itself then — especially for here-and-now applications — must be streamlined and focused on the fundamental pieces of the task. Although you *can* provide a near-infinity of details just to get a single task done, here's a word to the wise: Don't. You need to extract the essence of each task; focus on the details that really make a difference.

Although the travel app I describe at the beginning of this chapter is a good example, a counterexample might also be helpful.

The other night, my wife and I were standing with some friends inside the lobby of a movie theater, trying to decide where to go to grab some dinner. It was cold (at least by California standards), but we wanted to walk to the restaurant from the theater. We had two iPhones going, switching from application to application, trying to get enough information to make a decision. None of the applications gave us what we really needed — restaurants ranked by distance and type, with reviews and directions.

One of the applications was a great example of how to frustrate the user. It allowed you to select a restaurant by distance and cuisine. After you selected the distance, it gave you a list of cuisines. So far, so good. But the cuisine list was not context based; when I tapped *Ethiopian,* all I got was a blank screen. Very annoying! I took it off my iPhone then and there — I don't want an application that makes me work only to receive nothing in return. Your users won't either.

Every piece of a good application is not merely important to the task, but important to *where you are in the task*. For example, if I'm trying to decide how to get to central London from Heathrow, don't give me detailed information about the Tube until I need it.

In some ways, what you leave out is as important to the user experience as what you include.

That doesn't mean that your applications shouldn't make connections that ought to be made. One aspect of a compelling user experience is that all the pieces of an application work together to tell a story. If the tasks in your application are completely unconnected, perhaps they should be separate applications.

An application such as MobileTravel411 is aimed at people who may not know anything about their destination. If the application informs them that one reason to take the Heathrow Express is that it offers convenient Tube access on arrival in London, the users then have a bite-sized piece of valuable information about how to get around London when they're in the city. Save the train routes for when they're in the station.

Finally, limiting the focus to a single task also enables you to leave behind some iPhone (or any other mobile device) constraints such as limited screen size; in this respect, the limitations of the iPhone can guide you to a better application design.

It works the way I do

Great applications are based on the way people — users — think and work. When you make your application a natural extension of the user's world, it makes the application much easier and more pleasant to use — and to master.

Your users already have a mental model that describes the task your software is enabling. The users also have their own mental models of how the device works. At the levels of both content and user interface, your application must be consistent with these models if you want to create a superb user experience (which in turn creates loyalty — to your application).

The user interface in MobileTravel411 was based on how people divide the experience of traveling. Here are typical categories:

- Foreign currency — how much it really costs and what's the best way to convert money and buy things abroad
- Getting to and from the airport with maximum efficiency and minimum hassle
- Getting around a city, especially an unfamiliar one
- Finding any special events happening while you're in the city
- Handling traveler's tasks — such as making phone calls, tipping, or finding a bank or ATM — with aplomb
- Checking the current weather and the forecast
- Staying safe in unfamiliar territory — places you shouldn't go, what to do if you get in trouble, and so on

This is only a partial list, of course. I get into this aspect of application design in more detail when I take you through the design of MobileTravel411.

I suppose there are other ways to divide the tasks, but anything much different would be ignoring the user's mental model — which would mean that the application would not meet some of the user's expectations. It would be less pleasant to use because it would impose an unfamiliar way of looking at things instead of building on the knowledge and experiences those users already have.

 When possible, model your application's objects and actions on objects and actions in the real world. For example, the iPhone has a set of iPod-style playback controls, tapping controls to make things happen, sliding on-off switches, and flicking through the data shown on Picker wheels. All of these are based on physical counterparts in the real world.

Your application's text should be based on the target user. For example, if your user isn't steeped in technical jargon, avoid it in the user interface.

This doesn't mean that you have to "dumb down" the application. Here are some guidelines:

- ✔ If your application is targeted for a set of users who already use (and expect) a certain kind of specialized language, then sure, you can use the jargon in your application. Just do your homework first and make sure that you use those terms *correctly*.

  For example, if your application is targeted at high-powered foreign-exchange traders, your application might use *pip* ("price interest point" — the smallest amount that a price can move, as when a stock price advances by one cent). In fact, a foreign-exchange trader expects to see price movement in pips, and not only *can* you, but you *should* use that term in your user interface.

- ✔ If your application requires that the user have a certain amount of specialized knowledge about a task in order to use the application, identify what that knowledge is upfront.

- ✔ If the user is an ordinary person with generalized knowledge, use ordinary language.

 Gear your application to your user's knowledge base. In effect, meet your users where they are; don't expect them to come to you.

Disappearing the Technology

Although a few people out there actually love the gadgetry associated with mobile (or any) technology, at the end of the day, when you want to get something done, you'd really like the technology to disappear. The device itself should become simply an integral part of the environment you're in.

If you base your application on how the user interacts and thinks about the world, designing a great user interface becomes a whole lot easier. But that doesn't mean there aren't a significant number of ways to still blow it.

The user interface — form following function

Don't underestimate the effect of the user interface on the people who are trying to use it. A bad user interface can make even a great application painful to use. If users can't quickly figure out how to use your application, or if the user interface is cluttered or obscure, they're likely to move on and probably complain loudly about the application to anyone who will listen.

Simplicity and ease of use are fundamental principles for all types of software, but in iPhone applications, they are critical. Why? iPhone OS users are probably in the middle of other things while they use your application.

The iPhone hardware and software are outstanding examples of form following function; the user interfaces of great applications follow that principle as well. In fact, even the iPhone's limitations (except for battery life) are a result of form following from the functional requirements of a mobile device user. Just think how the iPhone fulfills the following mobile device user wish list:

✔ Small footprint

✔ Thin

✔ Lightweight

✔ Self-contained — no need for an external keyboard or mouse

✔ Task-oriented

It's a pretty safe bet that part of the appeal of the iPhone to many people — especially to non-technical users (like most of my friends) — is aesthetic: The device is sleek, compact, and fun to use. But the aesthetics of an iPhone application aren't just about how beautiful your application is onscreen. Eye candy is all well and good, but how well does your user interface match its function — that is, do its job?

Operational consistency

As with the Macintosh, users have a general sense of how applications work on the iPhone. (The Windows OS has always been a bit less user friendly, if you ask a typical Mac user.) One of the early appeals of the Macintosh was how similarly all the applications worked. So Apple (no fools they) carried over this similarity into the iPhone as well. The resulting success story suggests the following word to the wise. . . .

A compelling iPhone user experience usually requires familiar iPhone interface components offering standard functionality, such as searching and navigating hierarchical sets of data. Use the iPhone standard behavior, gestures, and metaphors in standard ways. For example, users tap a button to make a selection and flick or drag to scroll a long list. iPhone users understand these gestures because the built-in applications utilize them *consistently*. Fortunately, staying consistent is easy to do on the iPhone; the frameworks at your disposal have that behavior built in. This is not to say that you should never extend the interface, especially if you're blazing new trails or creating a new game. For example, if you want to "unlock" something in your user interface, why not use a pinch-open gesture, which adheres to the spirit if not the letter of the law, spelling out how to use the "standard" gesture?

Making it obvious

Although simplicity is a definite design principle, great applications are really about being easily understandable to the target user. If I'm designing a travel application, it has to be simple enough for even an inexperienced traveler to use. But if I'm designing an application for foreign-exchange trading, I don't have to make it simple enough for someone with no trading experience to understand.

Keep the following points in mind as you develop an application:

- The main function of a good application is immediately apparent and accessible to the users it's intended for.

- The standard interface components also give cues to the users. Users know, for example, to touch buttons and select items from table views (as in the contact application).

- You can't assume that users are so excited about your application that they're willing to invest lots of time in figuring it out.

Early Macintosh developers were aware of these principles. They knew that users expected that they could rip off the shrink-wrap, put a floppy disk in the machine (these were *really* early Macintosh developers), and do at least something productive immediately. The technology has changed since then; user attitudes, by and large, haven't.

Engaging the user

While I'm on the subject of users, here are two more aspects of a compelling application: direct manipulation and immediate feedback.

- **Direct manipulation makes people feel more in control.** On the desktop, that meant a keyboard and mouse; on the iPhone, the Multi-Touch interface serves the same purpose. In fact, using fingers gives a user a more immediate sense of control; there's no intermediary (such as a mouse) between the user and the object onscreen. To make this effect happen in your application, keep your onscreen objects visible while the user manipulates them, for example.

- **Immediate feedback keeps the users engaged.** Great applications respond to every user action with some visible feedback — such as highlighting list items briefly when users tap them.

Because of the limitations imposed by using fingers — especially somewhat pudgy fingers — applications need to be very forgiving. For example, although the iPhone doesn't pester the user to confirm every action, it also won't let the user perform potentially destructive, non-recoverable actions (such as deleting all contacts or restarting a game) without asking, "Are you sure?" Your application should also allow the user to easily stop a task that's taking too long to complete.

Notice how the iPhone uses animation to provide feedback. (I especially like the flipping transitions in the Weather application when I touch the Info button.) But keep it simple; excessive or pointless animation interferes with the application flow, reduces performance, and can really annoy the user.

An app may have an infrastructure the size of Texas behind it, but to the user, it should always appear to be just me and my app.

How the iPhone Makes This All Real

The iPhone's unique software and hardware allow you to create an application that enables the user to do something that may not be practical — or even possible — with a laptop computer. Although the iPhone is a smaller, mobile personal computer, it isn't a replacement for one. It isn't intended to produce documents, proposals, or research. The iPhone has the capability to be an extension of the user, seamlessly integrated into his or her everyday life, and able to accomplish a singly focused task, or step in a series of tasks, in real time, based on where he or she is.

Device-guided design

Although the enormous capabilities of the iPhone make it possible to deliver the compelling user experience your user craves, you must take into account the limitations of the device as well. Keeping the two in balance is *device-guided design.* The next two sections describe both the features and limitations of the iPhone — and how to take them into account as you plan and develop an application. But understanding these constraints can also inspire you to create some really innovative applications. After a closer look at device-guided design, I come back to what makes a compelling user experience.

Exploiting the features

The first thing that comes to mind when you think of an iPhone is mobility. This mobility really has two (separate and distinct) aspects to it:

 ✔ The device is small and unobtrusive enough to have it with you wherever you go. (Pretty obvious, that one.)

 ✔ The device is easily connected. (Look, Ma! No wires.) This fact leads to an interesting corollary: You don't need to have everything stored on the device. All you really need to know how to do is how to jump on the Internet and grab what you need from there.

I'm going to be talking a lot about mobility and what it means for developing iPhone apps during the course of this book, but a number of other hardware and software features built in to the device also enable a rich and compelling experience for the user. Clearly, one of the keys to creating a great application involves taking advantage of precisely these features. In the following sections, I offer a brief survey of the various device features you may want to use. I get into many of them in the book, but let this begin to pique your imagination.

Background processing and posting local notifications

While iOS (the iPhone operating system) doesn't have true multitasking (in fact, unless you have multiple cores or CPUs, no device has it), it does have instant-on task switching that reduces application startup and makes it easier to continue right where you left off. What you also have is the ability for certain kinds of applications to process events in the background. Such applications include

 ✔ **Audio:** The application plays audio in the background.

 ✔ **Location:** The application processes *location events* (information the iOS sends you about changes in location) in the background.

 ✔ **VoIP:** The application provides the ability for the user to make Voice over Internet Protocol calls — turning a standard Internet connection into a way to place phone calls.

What you can also do is post local notifications as a way to get a user's attention when important events happen. (This is especially relevant to Location Services, which I cover in greater detail in Chapter 18.) For example, a GPS navigation-type application running in the background can use local notifications to alert the user when it's time to make a turn. Applications can also schedule the delivery of local notifications for a future date and time and have those notifications delivered even if the application isn't running.

Accessing the Internet

The ability to access Web sites and servers on the Internet allows you to create applications that can provide real-time information to the user. It can tell me, for example, that the next tour at the Tate Modern is at 3 p.m. This kind of access also allows you, as the developer, to go beyond the limited memory and processing power of the device and access large amounts of data stored on servers, or even offload the processing. I don't need all the information for every city in the world stored on my iPhone or have to strain the poor CPU to compute the best way to get someplace on the Tube. I can send the request to a server and have it do all that work.

This is *client-server computing* — a well-established software architecture where the client provides a way to make requests to a server on a network that's just waiting for the opportunity to do something. A Web browser is an example of a client accessing information from other Web sites that act as servers.

Address Book and Contacts

Your application can access the user's contacts on the phone and display that information in a different way, or use it as information in your application. As a user of the MobileTravel411 application, for example, you could enter the name and address of your hotel, and the application would file it in your Contacts database. That way, you have ready access to the hotel address — not only from MobileTravel411, but also from your phone and other applications. Then when you arrive at Paddington Station, the application can retrieve the address from Contacts and display directions for you. What's more, you can also present standard system interfaces for picking and creating contacts in the Address Book.

Calendar Events

If you can leverage the information stored in the Address Book and Contacts databases, it stands to reason you can do the same thing with the Calendar application. You can remind a user when they need to leave for the airport, for example, or create calendar events based on what's happening this week in London. These events show up in the Calendar application and in other applications that support that framework.

Maps and Location

The iPhone OS and hardware allow a developer to determine the device's current location, or even to be notified when that location changes. As people move, it may make sense for your application to tailor itself to where the user is moment by moment.

There are already plenty of iPhone applications that use location information to tell you where the nearest coffeehouse is, or even where your friends are. The MobileTravel411 application uses this information to tell you the nearest Tube stop and give you directions to your hotel.

When you know the user's location, you can even put it on a map, along with other places he or she may be interested in. In Chapter 17, I show you how easy that really is.

An application that provides these kinds of services can also run in the background, and what's more, because using the GPS chip can be a significant drain on the battery, you have access to services that consume less battery power — including a location-monitoring service that tracks significant changes by using only cellular information and the ability to define arbitrary regions and detect boundary crossings into or out of those regions.

Camera and Photo Library

Your application can also access the pictures stored on the user's phone — and by "access" I mean not only display them, but also use or even modify them. The Photos application, for example, lets you add a photo to a contact, and several applications enable you to edit your photos on the iPhone itself. You can also incorporate the standard system interface to actually use the camera as well.

Even more advanced use of the camera is supported. One of the most interesting is *augmented reality,* where you're looking through the iPhone camera at the real world that has computer-generated graphics superimposed. (Actually it's much easier — and cooler — than it sounds.)

Playing audio and video

The iPhone OS makes it easy to play and include audio and video in your application. You can play sound effects or take advantage of the multichannel audio and mixing capabilities available to you. You can also create your own music player that has access to all the audio contents of the user's iPod Library. You can also play back many standard movie file formats, configure the aspect ratio, and specify whether or not controls are displayed. This means that your application can not only use the iPhone as a media player, but also use and control pre-rendered content. And you now can even record and edit HD videos. Keep in mind that if iMovie can do it, you probably can do it, too.

Let the games begin!

Copy, Cut, and Paste operations — and an Edit menu, to boot!

The iPhone OS supports Copy, Cut, and Paste operations within and between applications. It also provides a context-sensitive Edit menu that can display the Copy, Cut, Paste, Select, Select All, and Delete system commands. So that means that while on the iPhone each application is generally expected to play only in its own sandbox, you do actually have ways to send small amounts of data between applications.

Phone, messages (SMS), and mail

It's easy for you to write applications that reduce the degree of separation between your user and their friends (okay, also your business colleagues). Your application can present the standard system interfaces for composing and sending e-mail or SMS messages.

Your app can also interact with phone-based information on devices that have a cellular radio to get information about a user's cellular service provider and be notified when cellular call events occur. You can even access information about active cellular telephone calls, and access cellular service provider information from the user's SIM card. This makes things like VoIP (Voice over Internet Protocol) applications a reality, because you can now do the same things you can do with the cell phone — talk to someone and use the applications on the iPhone at the same time.

User interaction event handling

User interaction events inform your application of user actions — such as when a user touches a view, tilts the device, or presses a button on the headset.

People use their fingers, rather than a mouse, to select and manipulate objects on the iPhone screen. The moves that do the work, called *gestures,* give the user a heightened sense of control and intimacy with the device. There is a set of standard gestures — taps, pinch-close and pinch-open, flicks, and drags — that are used in the applications supplied with the iPhone.

 I suggest strongly that you use *only* the standard gestures in your application. Even so, the iPhone's gesture-recognition hardware and software allow you to go beyond standard gestures when appropriate. Because you can monitor the movement of each finger to detect gestures, you can create your own, but use that capability sparingly — only when it's undoubtedly the right thing to do in your application.

But in addition to touches, the user can communicate with the device in another way. The iPhone's new 3-axis gyroscope in combination with the existing accelerometer gives the device what's called *six-axis motion sensing* (x, y, z for both the accelerometer and gyroscope). You can work with both raw accelerometer and gyroscope data, and processed device-motion data (which includes accelerometer, rotation-rate, and altitude measurements). Such data is the basis for many games, as well as the shake-to-undo feature of the iPhone.

Hardware accessories

You can also create new hardware accessories and communicate with them either by a physical connection or Bluetooth. For example, there's a dongle that contains a transmitter you can use to broadcast audio from an iPhone to an FM radio.

Embracing the limitations

The preceding sections make it pretty clear that the iPhone is positively loaded with features you can leverage for use in your own application. All's not *completely* right with the world, however. Along with all those features, the iPhone *does* have some limitations. The key to successful applications — and to not making yourself too crazy — comes in understanding those limitations, living (and programming) within them, and even growing to love them. (It can be done. Honest.) These constraints help you understand the kinds of applications that are right for this device.

Often, it's likely that if you *can't* do something (easily, anyway) because of the iPhone's limitations, then maybe you shouldn't.

So find how to live with and embrace some facts of iPhone life:

- ✔ The small screen
- ✔ Users with fat fingers (me included)
- ✔ Limited computer power, memory, and battery life

The next sections can help get you get closer to this state of enlightenment.

Living with the small screen

While the iPhone's screen size and resolution allow you to deliver some amazing applications, it's still pretty small. Yet while the small screen limits what you can display on a single page, I have managed to do some mental jujutsu on myself to really think of it as a feature.

When your user interface is simple and direct, the user can understand it more easily. With fewer items in a small display, users can find what they want more quickly. A small screen forces you to ruthlessly eliminate clutter and keep your text concise and to the point (the way you like your books, right?).

Designing for fingers

Although the Multi-Touch interface is an iPhone feature, it brings with it limitations as well. First of all, fingers aren't as precise as a mouse pointer, which makes some operations difficult (text selection, for example). User-interface elements need to be large enough (Apple recommends that anything a user has to select or manipulate with a finger be a minimum of 44 x 44 pixels in size), and spaced far enough apart so that users' fingers can find their way around the interface comfortably.

You also can do only so much by using fingers. There are definitely a lot fewer possibilities when using fingers than the combination of multi-button mouse and keyboard.

Because it's so much easier to make a mistake when using just fingers, you also need to ensure that you implement a robust — yet unobtrusive — undo mechanism. You don't want to have your users confirm every action (it makes using the application tedious), but on the other hand, you don't want your application to let anybody mistakenly delete a page without asking, "Are you *sure* this is what you *really* want to do?" Lost work is worse than tedious.

Another issue around fingers is that the keyboard is not that finger-friendly. I admit it, using the iPhone keyboard is not up there on the list of things I really like about my iPhone. So instead of requiring the user to type some information, Apple suggests that you have a user select an item from a list. But on the other hand, the items in the list must be large enough to be easily selectable, which gets back to the first problem.

But again, like the small screen, this limitation can inspire (okay, may force) you to create a better application. To create a complete list of choices, for example, the application developer is forced to completely understand the context of (and be creative about) what the user is trying to accomplish. Having that depth of understanding then makes it possible to focus the application on the essential, eliminating what is unnecessary or distracting. It also serves to focus the user on the task at hand.

Limited computer power, memory, and battery life

As an application designer for the iPhone, you have several balancing acts to keep in mind:

- ✔ Although significant by the original Macintosh's standards, the computer power and amount of memory on the iPhone are limited.

- ✔ Although access to the Internet can mitigate the power and memory limitations by storing data and (sometimes) offloading processing to a server, those operations eat up the battery faster.

- ✔ Although the power-management system in the iPhone OS conserves power by shutting down any hardware features that aren't currently being used, a developer must manage the trade-off between all those busy features and shorter battery life. Any application that takes advantage of Internet access by using Wi-Fi or the 3G network, core location, and a couple of accelerometers is going to eat up the batteries.

The iPhone OS is particularly unforgiving when it comes to memory usage. If you run out of memory, it will simply shut you down.

This just goes to show that not *all* limitations can be exploited as "features."

Why Develop iPhone Applications?

Because you can. Because it's time. And because it's fun. Developing my iPhone applications has been the most fun I've had in many years (don't tell my wife!). Here's what makes it so much fun (for me, anyway):

✔ **iPhone apps are usually bite-sized — small enough to get your head around.** A single developer — or one with a partner and maybe some graphics support — can do them. You don't need a 20-person project with endless procedures and processes and meetings to create something valuable.

✔ **The applications are crisp and clean, focusing on what the user wants to do at a particular time and/or place.** They're simple but not simplistic. This makes application design (and subsequent implementation) much easier — and faster.

✔ **The free iPhone Software Development Kit (SDK) makes development as easy as possible.** I reveal its splendors to you throughout this book.

If you can't stand waiting, you *could* go on to Chapter 3, register as an iPhone developer, and download the SDK . . . but (fair warning) jumping the gun leads to extra hassle. It's worth getting a handle on the ins and outs of iPhone application development beforehand.

The iPhone has three other advantages that are important to you as a developer:

✔ **The App Store.** Apple will list your application in the App Store, and take care of credit card processing, hosting, downloading, notifying users of updates, and all those things that most developers hate doing. Developers name their own prices for their creations; Apple gets 30 percent of the sales price, with the developer getting the rest.

✔ **Apple has an iPhone developer program.** To get your application into the store, you have to pay $99 to join the program. But that's it. There are none of the infamous "hidden charges" that you often encounter, especially when dealing with credit-card companies. I explain how to join the developer program in Chapter 3 and how to work with the App Store in Chapter 12.

✔ **It's a business tool.** The iPhone has become an acceptable business tool, in part because it has tight security, as well as support for Microsoft Exchange and Office. This happy state of affairs expands the possible audience for your application.

Examining the possibilities

Just as the iPhone can extend the reach of the user, the device possibilities and the development environment can extend your reach as a developer. Apple talks often about three different application styles:

- ✔ **Productivity applications use and manipulate information.** The MobileTravel411 application is an example.

- ✔ **Utility applications perform simple, highly defined tasks.** The Weather Application is an example.

- ✔ **Immersive applications are focused on delivering — and having the user interact with — content in a visually rich environment.** A game is a typical example of an immersive application.

Although these categories help you understand how Apple thinks about iPhone applications (at least publicly), don't let them get in the way of your creativity. You've probably heard ad nauseam about stepping outside the box. But hold on to your lunch; the iPhone box isn't even a box yet. So here's a more extreme metaphor: Try diving into the abyss and coming up with something really new.

The Sample Applications

When I started writing the first edition of this book (which seems like an eternity but was really only 18 months ago), I decided that what I *wasn't* going to do was take the easy way out. By that I mean create a cookbook where you discuss a feature and then show a little bit of code that implements it.

Instead, I wanted to use real-world applications to illustrate the concepts, explain the technology, and give you the kind of background in the static and runtime architecture that you need if you want to develop applications that really exploit the iPhone's feature set.

In Figure 1-2, you can see the first application that I show you how to develop — the one I thought about after I lost my iPhone for the first time. I realized that if anyone found it and wanted to return it, well, returning it wouldn't be easy. Sure, whoever found it could root around in my Contacts or Favorites and maybe call a few of them and ask if any of their friends had lost an iPhone. But to save them the work, I decided to create an application called *ReturnMeTo,* whose icon sat on the upper-left corner of the home screen and looked like something you would want to select if you had found this phone. It would show a phone number to call and create a very happy person at the receiving end of the call.

Originally, I thought I would simply create this application and then get on with the rest of the book. It turned out, however, that as I showed my friends the application, I got a lot of feedback and made some changes to it. I include those changes as well — because they give you insight to the iPhone application-development process. All my friends also told me that they'd love to have the application. (Hey, they're my friends, after all.)

To pay them back for using them as a test group, I uploaded it to the App Store — I show you how to do the same with yours, in detail, in Chapter 12.

Although that app has been eclipsed by some of the newer features rolled out as part of the new iPhone OS 4, it's still a worthwhile exercise. Before, during, or after you've completed it yourself, go to the App Store and download it to your phone to see what this book is about — showing you what you need to do to create your own apps and get them into the App Store.

After I go through developing ReturnMeTo, I take you through the design of MobileTravel411 in Chapter 13. Then I show you how to implement a subset of this application, iPhoneTravel411, which shows you how to use much of the technology that implements the functionality of the MobileTravel411 application. You find out how to use table views (like the ones you see in the Contacts, iPod, Mail, and Settings applications that come with the iPhone), access data on the Web, go out to and return from Web sites while staying in your application, store data in files, include data with your application, allow users to set preferences, and even how to resume your application where the user last left off. I even talk about localization and self-configuring controllers and models. (Don't worry; by the time you get there, you'll know exactly what they mean.)

Because mapping and working with the user's location is such a big part of mobile applications, I show you how easy it is to create custom maps that are tailored to the needs of the user based on what she's doing and where she is. These maps will have annotations (those pins you see in the Maps application that give you information about that location when you tap them), and the annotations will be draggable so the user can change their location and what it says when they are tapped, Finally I show you how to monitor a user's location in the background and notify him or her when they need to know something — even when your app isn't running!

I use real-world applications to show the concepts and give you the background on how things really work on the iPhone — the in-depth knowledge you really need to go beyond the simple "Hello World" apps and create those killer iPhone applications. So be prepared! There may be some places where you might want to say, "Get on with it," but — based on my experience — I'm giving you what you'll need to move from following recipes in a cookbook by rote to modifying and even creating your own recipes.

Figure 1-2:
ReturnMeTo
— please!

What's Next

I'm sure that you're raring to go now and just can't wait to download the Software Development Kit (SDK) from the iPhone Developer Web site. That's exactly what I did — and later was sorry that I didn't spend more time upfront understanding how applications work in the iPhone environment.

So I ask you to be patient. In the next chapter, I explain what goes on behind the screen, and then, I promise, it's off to the races.

Chapter 2

Looking Behind the Screen

*O*ne of the things that makes iPhone software development so appealing is the richness of the tools and frameworks provided in the Apple iPhone Software Development Kit (SDK) for iPhone Applications. The *frameworks* are especially important; each one is a distinct body of code that actually implements your application's generic functionality — frameworks give the application its basic way of working, in other words. This is especially true of one framework in particular: the UIKit framework, the heart of the user interface.

In this chapter, I lead you on a journey through most of the iPhone's user interface architecture — a mostly static view that explains what the various pieces are, what each does, and how they interact with each other. This will lay the groundwork for developing the ReturnMeTo application's user interface, which you get a chance to tackle in Chapter 5. After that's done — but before you start major coding — I take you on a similar tour of the iPhone application *runtime environment* — the dynamic view of all the pieces working together when, for example, the user launches your application or touches a button on the screen.

Using Frameworks

A framework is designed to easily integrate any of the code that gives your application its specific functionality — the code that runs your game or delivers the information that your user wants, for example. Frameworks are therefore similar to software libraries, but with an added twist. Frameworks also implement a program's flow of control, unlike in a software library where it's dictated by the programmer. So instead of the programmer deciding in what order things happen — what messages are sent to what objects and in what order when an application launches, or what messages are sent to what objects in what order when a user touches a button on the screen — all of that is already a part of the framework and doesn't need to be specified by the programmer.

When you use a framework, you give your application a ready-made set of basic functions; you've told it, "Here's how to act like an application." With the framework in place, all you need to do is add the application's specific functionality you want — the content and the controls and views that enable the user to access and use that content — to the frameworks.

The frameworks and the iPhone OS provide some pretty complex functionality, such as

✔ Launching the application and displaying a window on the screen

✔ Displaying controls on the screen and responding to a user action — changing a toggle switch, for example, or scrolling a view, like the list of your contacts

✔ Accessing sites on the Internet, not just through a browser, but from within your own program

✔ Managing user preferences

✔ Playing sounds and movies

✔ The list goes on — you get the picture.

 Some developers talk in terms of "using a framework." I think about the matter differently: You don't use frameworks so much as they use you. You provide the functions that the framework accesses; it needs your code in order to become an application that does something other than start up, display a blank window, and then end. This perspective makes figuring out how to work with a framework much easier. (For one thing, it lets the programmer know where he or she is essential.)

If this seems too good to be true, well, okay, it is — all that complexity (and convenience) comes at a cost. It can be really difficult to get your head around the whole thing and know exactly where (and how) to add your application's functionality to that supplied by the framework. That's where *design patterns* come in. Understanding the design patterns behind the frameworks gives you a way of thinking about a framework — especially UIKit — that doesn't make your head explode.

Using Design Patterns

A major theme of this chapter is the fact that, when it comes to iPhone app development, the UIKit framework does a lot of the heavy lifting for you. That's all well and good, but it's a little more complicated than that: The framework is designed around certain programming paradigms, also known as *design patterns*. The design pattern is a model that your own code must be consistent with.

To understand how to take best advantage of the power of the framework — or (better put) how the framework objects want to use *you* best — you need to understand design patterns. If you don't understand them or if you try to work around them because you're sure that you have a "better" way of doing things, it will actually make your job much more difficult. (Developing software can be hard enough, so making your job more difficult is definitely something you want to avoid.) Getting a handle on the basic design patterns used (and expected by) the framework helps you develop applications that make the best use of the frameworks. This means the least amount of work in the shortest amount of time.

The iPhone design patterns can help you to understand not only how to structure your code, but also how the framework itself is structured. They describe relationships and interactions between classes or objects, as well as how responsibilities should be distributed amongst classes so the iPhone does what you want it to do.

The common definition of a design pattern is "a solution to a problem in a context." (Uh, guys, that's not too helpful.) At that level of abstraction, the concept gets fuzzy and ambiguous. So here's how I use the term throughout this book:

> In programming terms, a *design pattern* is a commonly used template that gives you a consistent way to get a particular task done.

There are five basic design patterns you need to be comfortable with:

- ✔ Model-View-Controller (MVC)
- ✔ Delegation
- ✔ Block Object
- ✔ Target-Action
- ✔ Managed Memory Model

I start with the Model-View-Controller design pattern, which is the key to understanding how an iPhone application works. I defer the discussion of the next three until after you get the MVC under your belt. As for the Managed Memory Model pattern, I explain that one in Chapter 6.

There's actually a sixth basic design pattern out there: threads and concurrent programming. This pattern enables you to execute tasks concurrently (including, in iPhone OS 4.0, the use of Grand Central Dispatch) and is way beyond the scope of this book.

The Model-View-Controller (MVC) pattern

The iPhone frameworks are *object-oriented*. The easiest way to understand what that really means is to think about a team. The work that needs to get done is divided up and assigned to individual team members (objects). Every member of a team has a job and works with other team members to get things done. What's more, a good team doesn't butt in on what other members are doing — just like how objects in object-oriented programming spend their time taking care of business and not caring what the object in the virtual cubicle next door is doing.

Object-oriented programming was originally developed to make code more maintainable, reusable, extensible, and understandable (what a concept!) by tucking all the functionality behind well-defined interfaces — the actual details of how something works (as well as its data) remains hidden. This makes modifying and extending an application much easier.

Great — so far — but a pesky question still plagues programmers:

Exactly how do you decide on the objects and what each one does?

Sometimes the answer to that question is pretty easy — just use the real world as a model (Eureka!). In the MobileTravel411 application that serves as an example later in this book, two of the classes I use are `Airport` and `Currency`. But when it comes to a generic program structure, how *do* you decide what the objects should be? That may not be so obvious.

The MVC pattern is a well-established way to group application functions into objects. Variations of it have been around at least since the early days of Smalltalk, one of the very first object-oriented languages. The MVC is a high-level pattern — it addresses the architecture of an application and classifies objects according to the (very) general roles they play in an application.

The MVC pattern creates, in effect, a miniature universe for the application, populated with three kinds of objects. It also specifies roles and responsibilities for all three objects and specifies the way they're supposed to interact with each other. To make things more concrete (that is, to keep your head from exploding), imagine a big, beautiful, 60-inch flat-screen TV. Here's the gist:

- **Model objects:** These objects together comprise the content engine of your application. They contain the application's data and logic — making your application more than just a pretty face. In the Mobile Travel411 application, the model knows the various ways to get from Heathrow Airport to London as well as some logic to decide the best alternative based on time of day, price, and some other considerations.

  You can think of the *model* (which may be one object or several that interact) as a particular television program. One that, quite frankly, doesn't give a hoot about what TV set it is being shown on.

  In fact, the model shouldn't give a hoot. Even though it owns its data, it should have no connection at all to the user interface and should be blissfully ignorant about what's being done with its data.

- **View objects:** These objects display things on the screen and respond to user actions. Pretty much anything you can see is a kind of view object — the window and all the controls, for example. Your views know how to display information that they have gotten from the model object, and how to get any input from the user the model may need. But the view itself should know nothing about the model. It may handle a request to tell the user the fastest way to London, but it doesn't bother itself with what that request means. It may display the different ways to get to London, although it doesn't care about the content options it displays for you.

  You can think of the *view* as a television screen that doesn't care about what program it's showing or what channel you just selected.

  The UIKit framework provides many different kinds of views, as you find out later on in this chapter in the "Working with Windows and Views" section.

  If the view knows nothing about the model, and the model knows nothing about the view, how do you get data and other notifications to pass from one to the other? To get that conversation started ("Model: I've just updated my data." View: "Hey, give me something to display," for example), you need the third element in the MVC triumvirate, the controller.

- **Controller objects:** These objects connect the application's view objects to its model objects. They supply the view objects with what they need to display (getting it from the model), and also provide the model with user input from the view.

You can think of the *controller* as the circuitry that pulls the show off of the cable, and sends it to the screen, or requests a particular pay-per-view show.

The MVC in action

Imagine that an iPhone user is at Heathrow Airport, and he or she starts the handy MobileTravel411 application mentioned so often in these pages. The view displays his or her location as "Heathrow Airport." The user may tap a button that requests the best way to get into London. The controller interprets that request and tells the model what it needs to do by sending a message to the appropriate method in the model object with the necessary parameters. The model computes a list of alternatives (taxi, bus, train), and the controller then delivers that information to the view, which promptly displays it. If the user selects the train option, for example, that information is then sent to the model, which then sends back the details of the train.

All this is illustrated in Figure 2-1.

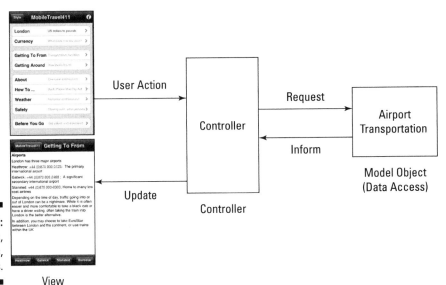

Figure 2-1:
Models,
controllers,
and views.

When you think about your application in terms of model, view, and controller objects, the UIKit framework starts to make sense. It also begins to lift the fog from where at least part of your application-specific behavior needs to go. Before I get more into that, however, you need to know a little more about the classes provided to you by the UIKit that implement the MVC design pattern — windows, views, and view controllers.

Working with Windows and Views

After an application is launched, it's going to be the only application running on the system — aside from the operating system software, of course. iPhone applications have only a single window, so you won't find separate document windows for displaying content. Instead, everything is displayed in that single window, and your application interface takes over the entire screen. When your application is running, it's all the user is doing with the iPhone.

Looking out the window

The single window you see displayed on the iPhone is an instance of the UIWindow class. This window is created at launch time, either programmatically by you or automatically by UIKit loading it from a *nib* file — a special file that contains instant objects that are reconstituted at runtime (You find out more about nib files, starting in Chapter 6). You then add views to the window. In general, after you create the window object (that is, if you create it instead of having it done for you), you never really have to think about it again.

An iPhone window cannot be closed or manipulated directly by the user. It's your application that programmatically manages the window.

Although your application never creates more than one window at a time, the iPhone OS does use additional windows on top of your window. The system status bar is one example. You can also display alerts on top of your window by using the supplied alert views.

Figure 2-2 shows the window layout on the iPhone for the MobileTravel411 application.

Status bar ———→

Navigation bar ———→

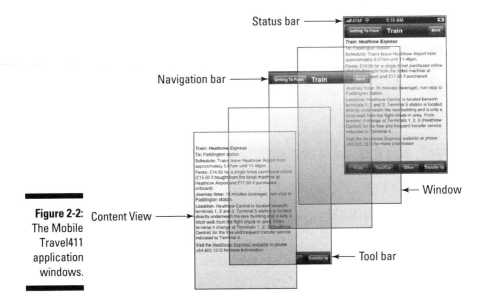

←——— Window

Figure 2-2:
The Mobile
Travel411
application
windows.

Content View ———→

←——— Tool bar

Admiring the view

In an iPhone app world, view objects are responsible for the view functionality in the Model-View-Controller architecture.

A view is a rectangular area on the screen (on top of a window). I often refer to the *content view,* that portion of data and controls that appears between the upper and lower bars shown in Figure 2-2.

In the UIKit framework, windows are really a special kind of view, but for purposes of this discussion, I'm going to be talking about views that sit on top of the window.

As you will see, there are two ways you need to think about views. From the user perspective, the views sit on top of each other. From a programming perspective, however, the views that are on top of the windows visually are really subviews inside the window view. I explain that more later on.

What views do

Views are the main way for your application to interact with a user. This interaction happens in two ways:

- ✔ **Views display content.** For example, making drawing and animation happen onscreen.

  In essence, the view object displays the data from the model object.

✔ **Views handle touch events.** They respond when the user touches a button, for example.

Handling touch events is part of a *responder chain* (a special logical sequence detailed in Chapter 6).

The view hierarchy

Views and subviews create a view hierarchy. There are two ways of looking at it (no pun intended this time): visually (how the user perceives it) and programmatically (how you create it). You must be clear about the differences, or you'll find yourself in a state of confusion that resembles Times Square on New Year's Eve.

Looking at it visually, the window is at the base of this hierarchy with a *content view* on top of it (a transparent view that fills the window's Content rectangle). The content view displays information and also allows the user to interact with the application, using (preferably standard) user interface items such as text fields, buttons, toolbars, and tables.

In your program, that relationship is different. The content view is added to the window view as a *subview*.

✔ Views added to the content view become *subviews* of it.

✔ Views added to the content view become the *superviews* of any views added to them.

✔ A view can have one (and only one) superview and zero or more subviews.

It seems counterintuitive, but a subview is displayed *on top of* its parent view (that is, on top of its superview). Think about this relationship as containment: a superview *contains* its subviews. Figure 2-3 shows an example of a view hierarchy.

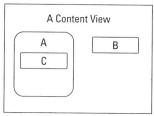

The visual hierarchy
... translates to a structural one:

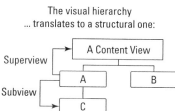

Figure 2-3: The view hierarchy is both visual and structural.

Controls — such as buttons, text fields, and the like — are really view subclasses that become subviews. So are any other display areas you may specify. The view must manage its subviews, as well as resize itself with respect to its superviews. Fortunately, much of what the view must do is already coded for you. The UIKit framework supplies the code that defines view behavior.

The view hierarchy plays a key role in both drawing and event handling. When a window is sent a message to display itself, the window asks its subview to render itself first. If that view has a subview, it asks *its* subview to render itself first, going down the structural hierarchy (or up the visual structure) until the last subview is reached. It then renders itself and returns to its caller, which renders itself, and so on.

You create or modify a view hierarchy whenever you add a view to another view, either programmatically or with the help of the Interface Builder. (You find more about Interface Builder in Chapter 5.) The UIKit framework automatically handles all the relationships associated with the view hierarchy.

I pretty much glossed over this visual versus programmatic view hierarchy stuff when I started developing my applications — making it really difficult to get a handle on what was going on. Don't make the same mistake I did.

The kinds of views you use

The UIView class defines the basic properties of a view, and you may be able to use it as is — like I do in the ReturnMeTo application later in this book — by simply adding some controls.

The framework also provides you with a number of other views that are subclassed from UIView. These views implement the kinds of things that you, as a developer, need to do on a regular basis.

It's important to use the view objects that are part of the UIKit framework. When you use an object such as a UISlider or UIButton, your slider or button behaves just like a slider or button in any other iPhone application. This enables the consistency in appearance and behavior across applications that users expect. (For more on how this kind of consistency is one of the characteristics of a great application, see Chapter 1.)

Container views

Container views are a technical (Apple) term for content views that do more than just lie there on the screen and display your controls and other content.

The UIScrollView class, for example, adds scrolling without you having to do any work.

UITableView inherits this scrolling capability from UIScrollView and adds the ability to display lists and respond to the selections of an item in that list. Think of the Contacts application (and a host of others). UITableView is one of the primary navigation views on the iPhone; you work a lot with table views, starting with Chapter 14.

Another container view, the UIToolbar class, contains buttonlike controls — and you find those everywhere on the iPhone. In Mail, for example, you touch an icon in the bottom toolbar to respond to an e-mail.

Controls

Controls are the fingertip-friendly graphics you see extensively used in a typical application's user interface. Controls are actually subclasses of the UIControl superclass, a subclass of the UIView class. They include touchable items like buttons, text fields, sliders, and switches, as well as text fields in which you enter data.

Controls make heavy use of the Target-Action design pattern, which I get to soon. (I talk more about controls and how they fit in to the Target-Action pattern in Chapter 11.)

Display views

Think of display views as controls that look good, but don't really do anything except, well, look good. These views include UIImageView, UILabel (which I use in Chapter 5 to display the ReturnMeTo application's phone number), UIProgressView, and UIActivityIndicatorView.

Text and Web views

Text and *Web views* provide a way to display formatted text in your application. The UITextView class supports the display and editing of multiple lines of text in a scrollable area. The UIWebView class provides a way to display HTML content. These views can be used as the content view, or can be used in the same way as a display view (see the preceding "Display views" section), as a subview of a content view. I use a UIWebView in Chapter 11 to allow someone who has found an iPhone to call the owner's number by simply tapping it. UIWebView also is the primary way to include graphics and formatted text in text display views. (I use them when I show you how to develop iPhoneTravel411.)

Alert views and action sheets

Alert views and *action sheets* present a message to the user, along with buttons that allow the user to respond to the message. Alert views and action sheets are similar in function but look and behave differently. For example, the UIAlertView class displays a blue alert box that pops up on the screen, and the UIActionSheet class displays a box that slides in from the bottom of the screen.

Navigation views

Tab bars and *navigation bars* work in conjunction with view controllers to provide tools for navigating in your application. Normally, you don't need to create a UITabBar or UINavigationBar directly — it's easier to use Interface Builder or configure these views through a tab bar or navigation bar controller.

The window

A *window* provides a surface for drawing content and is the root container for all other views.

There's typically only one window per application.

Controlling View Controllers

View controllers implement the controller component of the Model-View-Controller design pattern. These controller objects contain the code that connects the application's view objects to its model objects. They provide the data to the view. Whenever the view needs to display something, the view controller goes out and gets what the view needs from the model. Similarly, view controllers respond to controls in your content view and may do things like tell the model to update its data (when the user adds or changes text in a text field, for example), or compute something (the current value of, say, your U.S. dollars in British pounds), or change the view being displayed (like when the user hits the detail disclosure button on the iPod application to find out more about a song).

As I describe in "The Target-Action pattern" section later in this chapter, a view controller is often the (target) object that responds to the on-screen controls. The Target-Action mechanism is what enables the view controller to be aware of any changes in the view, which can then be transmitted to the model. For example, the user may decide — after looking at the Heathrow Express option — that he or she has too much luggage (or is too upscale) to take the train, and opts for a taxi or rental car instead.

Figure 2-4 shows what happens when the user taps the Taxi/Car tab in the MobileTravel411 application to request information about taking a cab or renting a car to get to London.

1. A message is sent to that view's view controller to handle the request.

2. The view controller's method interacts with a model object.

3. The model object processes the request from the user for information on a taxi/car from Heathrow to London.

4. The model object sends the data back to the view controller.

5. The view controller creates a new view to present the information.

View controllers have other vital iPhone responsibilities as well, such as:

✔ Managing a set of views — including creating them, or flushing them from memory during low-memory situations.

✔ Responding to a change in the device's orientation — say, landscape to portrait — by resizing the managed views to match the new orientation.

✔ Creating *modal* views that require the user to do something (touch the Yes button, for example) before returning to the application.

You would use a modal view to ensure that the user has paid attention to the implications of an action (for example, "Are you *sure* you want to delete all your contacts?").

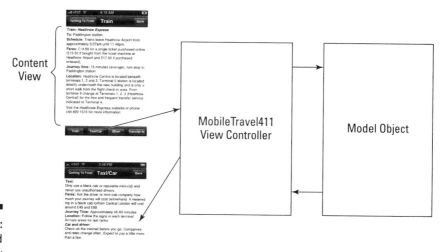

Content View

MobileTravel411 View Controller

Model Object

Figure 2-4:
The world of the view controller.

View controllers are also typically the objects that serve as delegates and data sources for table views (more about those in Chapter 14).

In addition to the base `UIViewController` class, `UIKit` includes subclasses such as `UITabBarController`, `UINavigationController`, `UITableViewController`, and `UIImagePickerController` to manage the tab bar, navigation bar, table views, and to access the camera and photo library.

Even if your application is a graphics application, you'll want to use a view controller just to manage a single view and auto-rotate it when the device's orientation changes.

What about the Model?

As you have seen (and will continue to discover), a lot of the functionality you need is already in the framework objects.

But when it comes to the model objects, for the most part, you're pretty much on your own. You're going to need to design and create Model objects to hold the data and carry out the logic. In my MobileTravel411 application, for example, you create an `Airport` object that knows the different ways to get into the city that it supports.

You may find classes in the framework that help you get the nuts and bolts of the model working. But the actual content and specific functionality is up to you. As for actually implementing model objects, I show you how to do that in Chapter 16.

Using naming conventions

When creating your own classes, it's a good idea to follow a couple of standard framework-naming conventions:

✔ Class names (such as `View`) should start with a capital letter.

✔ The names of methods (such as `viewDidLoad`) should start with a lowercase letter.

✔ The names of instance variables (such as `frame`) should start with a lowercase letter.

When you do it this way, it makes it easier to understand from the name what something actually is.

Adding Your Own Application's Behavior

Earlier in this chapter (by now it probably seems like a million years ago), I mention other design patterns used in addition to the Model-View-Controller (MVC) pattern. Three of these patterns — the Delegation pattern, the Target-Action pattern, and the Block Object pattern — along with the MVC pattern and subclassing, provide the mechanisms for you to add your application-specific behavior to the UIKit (and any other) framework.

Here are the ways to add behavior:

- ✔ I already talked about the first way to add behavior, and that's through model objects in the MVC pattern. Model objects contain the data and logic that make, well, your application.

- ✔ The second way — the way people traditionally think about adding behavior to an object-oriented program — is through *subclassing,* where you first create a new (sub) class that inherits behavior and instance variables from another (super) class and then add more behavior, instance variables, and *properties* (I explain properties in Chapter 7) to the mix until you come up with just what you want. The idea here is to start with something basic and then add to it — kind of like taking a deuce coupe (1932 Ford) and turning it into a hot rod. You'd subclass a view controller class, for example, to respond to controls.

- ✔ The third way to add behavior involves using the Delegation pattern, which allows you to customize an object's behavior without subclassing by basically forcing another object to do the first object's work for it. For example, the Delegation design pattern is used at application startup to invoke a method applicationDidFinishLaunching: that gives you a place to do your own application-specific initialization. All you do is add your code to the method.

- ✔ The fourth way to add behavior is by using block objects. The Block Object design pattern is similar to Delegation, but it's more *event driven* in that it allows you to create methods or functions that you can pass to other methods or functions that are executed as needed. For example, you might want to have some code that scrolls the view as necessary when the keyboard appears. You would pass that to a method that is invoked when the keyboard appears.

- ✔ The final way to add behavior involves the Target-Action design pattern, which allows your application to respond to an event. When a user touches a button, for example, you specify what method should be invoked to respond to the button touch. What's interesting about this pattern is that it also requires subclassing — usually a view controller (see the "Controlling View Controllers" section, earlier in this chapter) — in order to add the code to handle the event.

In the next few sections, I go into a little more detail about the Delegation, Block Object, and Target-Action patterns.

The Delegation pattern

Delegation is a pattern used extensively in the iPhone framework, so much so that it's very important to clearly understand. In fact, I have no problems telling you that, when you understand it, your life will be much easier. Until the light bulb went on for me, I sometimes felt like I was trying to make my way through one of those legendary London pea soup fogs.

As I said in the previous section, delegation is a way of customizing the behavior of an object without subclassing it. Instead, one object (a framework object) delegates the task of implementing one of its responsibilities to another object. You are using a behavior-rich object supplied by the framework as is and putting the code for program-specific behavior in a separate (delegate) object. When a request is made of the framework object, the method of the delegate that implements the program-specific behavior is automatically called.

For example, the `UIApplication` object handles most of the actual work needed to run the application. But, as you will see, it sends your application delegate the `application:didFinishLaunchingWithOptions:` message to give you an opportunity to restore the application's window and view to where it was when the user previously left off. You can also use this method to create objects that are unique to your application.

When a framework object has been designed to use delegates to implement certain behaviors, the behaviors it requires (or gives you the option to implement) are defined in a *protocol.*

Protocols define an interface that the delegate object implements. On the iPhone, protocols can be formal or informal, although I'm going to concentrate solely on the former because it includes support for things like type checking and runtime checking to see whether an object conforms to the protocol.

In a formal protocol, you usually don't have to implement all the methods; many are declared optional, meaning that you only have to implement the ones relevant to your application. Before it attempts to send a message to its delegate, the host object determines whether the delegate implements the method (via a `respondsToSelector:` message) to avoid the embarrassment of branching into nowhere if the method is not implemented.

You find out much more about delegation and the Delegation pattern when you develop the ReturnMeTo (and especially the iPhoneTravel411) applications in later chapters.

The Block Object pattern

Although delegation is extremely useful, it is not the only way to customize the behavior of a method or function.

Blocks are like traditional C functions in that they are small, self-contained units of code. They can be passed in as arguments of methods and functions and then used when they're needed to do some work. Like many programming topics, understanding block objects is easier when you use them. I show you how to use blocks in Chapter 8, where I also show you how you can use a block object to scroll a view when the user wants to enter text in a field that would be hidden by the keyboard.

With iOS 4, a number of methods and functions of the system frameworks are starting to take blocks as parameters:

- Completion handlers
- Notification handlers
- Error handlers
- Enumeration
- View animation and transitions
- Sorting

In Chapter 8, you use a block object to implement a *callback* — a method that's invoked based on an event (in this case, the user touching inside a text field which raises a keyboard). (In a sidebar, I also show you how to use a block in a view animation). Block objects also have a number of other uses, especially in Grand Central Dispatch and the NSOperationQueue class, the two recommended technologies for concurrent processing. But because concurrent processing is out of scope for this book (way out of scope in fact), I leave you to explore that use on your own.

The Target-Action pattern

The *Target-Action* pattern is used to let your application know that a user has done something. He or she may have tapped a button or entered some text, for example. The control — a button, say — sends a message (the *action* message) that you specify to the target you have selected to handle that particular action. The receiving object, or the *target,* is usually a view controller object.

If you wanted to start your car from your iPhone (not a bad idea if you've ever lived in some place like Minneapolis), you could display two buttons, Start and Heater. When you tapped Start, you could have used Interface Builder to specify that the target is the `CarController` object and that the method to invoke is `ignition`. Figure 2-5 shows the Target-Action mechanism in action. (If you're curious about `IBAction` and `(id) sender`, I explain what they are when I show you how to use the Target-Action pattern in your application in Chapter 11.)

Figure 2-5:
The Target-
Action
mechanism.

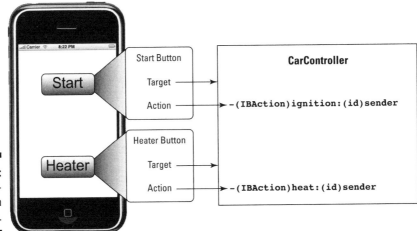

The Target-Action mechanism enables you to create a control object and tell it not only what object you want handling the event, but also the message to send. For example, if the user touches a Ring Bell button onscreen, I want to send a Ring Bell message to the view controller. But if the Wave Flag button on the same screen is touched, I want to be able to send the same view controller the Wave Flag message. If I couldn't specify the message, all buttons would have to send the same message. It would then make the coding more difficult and more complex because I would have to identify which button had sent the message and what to do in response, and that would make changing the user interface more work and more error prone.

As you'll soon discover when creating your application, you can set a control's action and target through the Interface Builder application. This ability allows you to specify what method in which object should respond to a control without having to write any code.

You can also change the target and action dynamically by sending the control or its cell `setTarget:` and `setAction:` messages.

For more on the Interface Builder application, check out Chapter 5.

Doing What When?

The `UIKit` framework provides a great deal of ready-made functionality, but the beauty of `UIKit` lies in the fact that — as this chapter makes clear — you can customize its behavior by using four distinct mechanisms.

- ✔ Subclassing
- ✔ Delegation
- ✔ Block Objects
- ✔ Target-Action

One of the challenges facing a new developer is to determine which of these mechanisms to use when. (That was certainly the case for me.) To ensure that you have an overall conceptual picture of the iPhone application architecture, check out the online Cheat Sheet for this book, where I give you a summary of which mechanisms are used when. (I wish I'd had this when I started developing my application — but at least you do now.) You can find the Cheat Sheet for this book at www.dummies.com/cheatsheet/iphone applicationdevelopment.

Whew!

Congratulations! You've just gone through the *Classic Comics* version of hundreds of pages of Apple documentation, reference manuals, and how-to guides.

Well, you still have a bit more to explore — for example, how all these pieces work together at runtime (details, details, . . .). But before that piece of the puzzle can make sense, you need to touch, feel, and get inside an application. As part of that process, I do a little demonstrating:

✔ I show you how to become a registered developer and download the SDK in Chapter 3.

✔ I show you how to use the tools in the SDK to create an application framework in Chapter 4 and how to build the user interface in Chapter 5.

✔ I finish the conversation on iPhone architecture in Chapter 6.

When you've had a stroll through those adventures, you'll know everything you need to know about how to create a user interface and add the functionality to make your application do what you promised the user it would do. (How's that for a plan?)

Chapter 3

Enlisting in the Developer Corps

*P*ersonally, I'm not much of a joiner. I like to keep a low profile and just get on with having fun doing what I do.

But if you want to develop software for the iPhone, you do have to get involved with (yet another) major corporation and its policies and procedures. Although the Apple iPhone Software Development Kit (SDK) is free, you do have to register as an iPhone developer first. That will give you access to all the documentation and other resources found on the iPhone Developer Web site. This whole ritual transforms you into a *Registered iPhone Developer.*

Becoming a Registered Developer is free, but there's a catch: If you actually want to run your application on your iPhone, as opposed to only on the Simulator that comes with the SDK, you have to join the developer program. Fortunately, membership only costs $99, but then again, you have no choice if you want your application to see the light of day on the iPhone. This is called joining the *iPhone Developer Program.*

In this chapter, I lead you through the process of becoming a Registered Developer; signing on to — and then exploring — the iPhone Dev Center Web site; downloading the SDK so you can start using it; and then (finally) joining the developer program.

What you see when you go through this process yourself may be slightly different from what you see here. Don't panic. It's because Apple changes the site from time to time.

Becoming a Registered iPhone Developer

Although just having to become a registered developer is annoying to some people, it doesn't help that the process itself can be a bit confusing as well. Fear not! Follow the steps, and I get you safely to the end of the road. (If you've already registered, skip to the next section where I show what the iPhone Dev Center has available as well as how to download the SDK.)

1. **Point your browser to `http://developer.apple.com/iphone`.**

   Doing so brings you to a page similar to the one shown in Figure 3-1. Apple does change this site occasionally, so when you get there, it may be a little different. You may be tempted by some of the links, but they only get you so far until you log in as a Registered Developer.

2. **Click the Register link in the top-right corner of the screen.**

   A page appears describing what you get after you register, as shown in Figure 3-2. At the top of this page, you can see a prominently displayed button that says Get Started, followed by the much beloved word — Free.

   In Figure 3-2, also notice at the very bottom of the page a Learn More about Our Developer Programs link. This is one way to become a registered developer. I show you another way later.

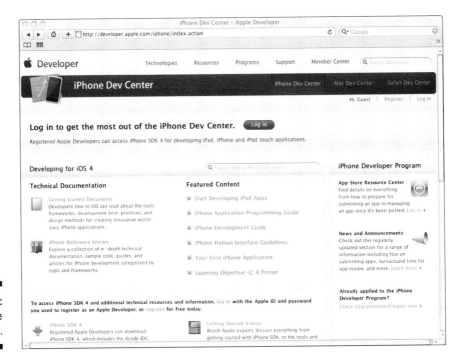

Figure 3-1:
The iPhone
Dev Center.

Figure 3-2:
Register as
an Apple
Developer.

3. Click the Get Started button.

You're taken to a page asking whether you want to create an Apple ID or use an existing one.

You can use your current Apple ID (the same one you use for iTunes or any other Apple Web site) or create an Apple ID and then log in.

- If you don't have an Apple ID, select Create an Apple ID and click Continue. Follow the yellow brick road so to speak and you'll soon have one.

- If you already have an Apple ID, select the Use an Existing Apple ID option and click Continue. You're taken to a screen where you can log in with your Apple ID and password. Logging in takes you to Step 4 with some of your information already filled out.

4. Fill out the forms. (What more can I say?)

After a few more pages, you'll be a registered developer and ready to roll.

5. You'll receive an e-mail from Apple Developer Support with a verification code. You need to enter that at some point in the process.

6. On the thank-you page, click Continue and on the Developer page that appears click iPhone under Dev Centers.

You now find yourself logged in to the iPhone Development Center. Here you can explore the resources available to you as well as join the iPhone Developer Program, as you can see in the right column of Figure 3-3.

You're now an officially registered iPhone developer. The next section starts, showing you what you can do with your new status.

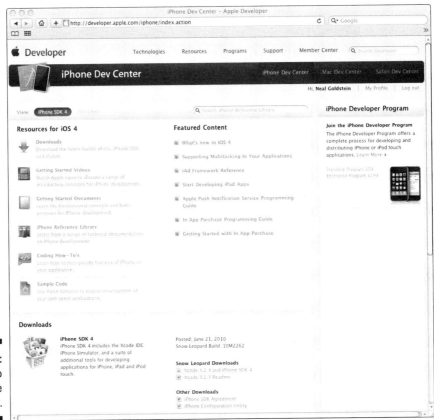

Figure 3-3:
Logged in to the iPhone Dev Center.

Exploring the iPhone Dev Center

I spend some time later in this section talking a little bit about some of the resources available to you in the iPhone Dev Center, but for the moment, let's focus on what you're *really* after. I'm talking about the iPhone SDK 4 download that you see in Figure 3-3 at the bottom of the iPhone Dev Center page.

The SDK includes a host of tools for you to develop your application. Here's a handy list to help you keep them all straight:

- **Xcode:** This refers to Apple's complete development environment, which integrates all these features: a code editor, build system, graphical debugger, and project management. (I introduce you to the code editor's features in more detail in Chapter 7.)

- **Frameworks:** The iPhone's multiple frameworks help make it easy to develop for. Creating an application can be thought of as simply adding your application-specific behavior to a framework. The frameworks do all the rest. For example, the UIKit framework provides fundamental code for building your application — the required application behavior, classes for windows, views (including those that display text and Web content), controls, and view controllers. (All the things I cover in Chapter 2, in other words.) The UIKit framework also provides standard interfaces to core location data, the user's contacts and photo library, accelerometer data, and the iPhone's built-in camera.

- **Interface Builder:** I use Interface Builder in Chapter 5 to build the user interface for the ReturnMeTo application. But Interface Builder is more than your run-of-the-mill program that builds graphical user interfaces. In Chapter 6, I show you how Xcode and Interface Builder work together to give you ways to build (and automatically create at runtime) the user interface — as well as to create objects that provide the infrastructure for your application.

- **iPhone Simulator:** The simulator allows you to debug your application and do some other testing on your Mac by simulating the iPhone. The Simulator will run most iPhone programs, but it doesn't support some hardware-dependent features. I give you a rundown on the Simulator in Chapter 4.

- **Instruments:** The Instruments application lets you measure your application while it's running on a real device. It gives you a number of performance metrics, including those for testing memory and network use. It will also work (in a limited way) on the iPhone Simulator, and you can test some aspects of your design there.

The iPhone Simulator doesn't emulate such real-life iPhone characteristics as CPU speed or memory throughput. If you want to understand how your application performs on the device from a user's perspective, you have to use the actual device in combination with the Instruments application.

Looking forward to using the SDK

The tools in the SDK support a development process that most people find comfortable. They allow you to rapidly get a user interface up and running to see what it actually looks like. You can add code a little at a time and then run it after each new addition to see how it works. I take you through this incremental process as I develop the ReturnMeTo application; for now, here's a bird's-eye view of iPhone application development, one step at a time:

1. **Start with Xcode.**

    Xcode provides several project templates that you can use to get you off to a fast start. (In Chapter 5, I do just that to get my user interface up and running quickly.)

2. **Design and create the user interface.**

    Interface Builder has graphic-design tools you can use to create your application's user interface. This saves a great deal of time and effort. It also reduces the amount of code you have to write by creating resource files that your application can then upload automatically.

    If you don't want to use Interface Builder, you can always build your user interface from scratch, creating each individual piece and linking them all together within your program itself. Sometimes Interface Builder is the best way to create onscreen elements; sometimes the hands-on approach works better.

3. **Write the code.**

    The Xcode editor provides several features that help you write code. I run through these features in Chapter 7.

4. **Build and run your application.**

    You build your application on your computer and run it in the iPhone Simulator application or (provided you've joined the Development Program) on your device.

5. **Test your application.**

    You'll want to test the functionality of your application as well as response time.

6. **Measure and tune your application's performance.**

    After you have a running application, you want to make sure that it makes optimal use of resources such as memory and CPU cycles.

7. **Do it all again until you're done.**

Resources on the iPhone Dev Center

You're not left on your own when it comes to the Seven-Step Plan for Creating Great iPhone Apps in the preceding section. After all, you have me to help you on the way — as well as a heap of information squirreled away in various corners of the iPhone Dev Center. I've found the following resources to be especially helpful:

- ✔ **Getting Started Videos:** There is a lot of stuff here. You may want to look through these after you're done with this book if you're looking for more information in a specific subject area.

- ✔ **Getting Started Documents:** Think of them as an introduction to the materials in the iPhone Reference Library. These give you an overview of iPhone development and best practices.

  If you then click the Guides link in the left column, you find a list of documents that include *Learning Objective-C: A Primer* (an overview of Objective-C), *Object-Oriented Programming with Objective-C*, and *The Objective-C 2.0 Programming Language* (the definitive guide).

  If you've never programmed in the Objective-C language, you can find some basic information in the iPhone Reference Library, as I mention earlier. But if you want to really master Objective-C as quickly (and pain-lessly) as possible, go get yourself a copy of *Objective-C For Dummies* by yours truly (Wiley). I explain everything you need to know in order to program in Objective-C, and assume that you have little or no knowledge of programming. (It does a great job, if I do say so myself.)

- ✔ **The iPhone Reference Library:** This is all the documentation you could ever want (except, of course, the answer to that one question you really need answered at 3 a.m., but that's the way it goes). To be honest, most of this stuff turns out to be really useful only *after* you have a good handle on what you're doing. As you go through this book, however, an easier way to access some of this documentation is through Xcode's Documentation window, which I show you in Chapter 7.

- ✔ **Coding How-To's:** These tend to be a lot more valuable when you already have something of a knowledge base.

- ✔ **Sample Code:** On the one hand, sample code of any kind is always valu-able. Most good developers look to these kinds of samples to get them-selves started. They take something that closely approximates what they want to do and then modify it until it does. When I started iPhone develop-ment, there were no books like this one; so much of what I learned came from looking at the samples and then making some changes to see how things worked. On the other hand, it can give you hours of (misguided) pleasure and can be quite the time waster and task avoider.

Downloading the SDK

Enough prep work. Time to do some downloading. Click the Xcode 3.2.3 and iPhone SDK 4 link under the Snow Leopard Downloads heading.

By the time you read this book, it may no longer be version 4. You should download the latest (non-beta, non-prelease) SDK. That way you get the most stable version to start with.

Under Featured Content, you can also find a number of helpful links. One you may want to look at is What's New in iOS 4. You can also safely ignore the Xcode 3.2.3 Readme under the Xcode 3.2.2 and iPhone SDK 4 link unless you're already familiar with Xcode and absolutely *have* to see what's new.

After clicking the link, you can watch the download in Safari's download window (which is only a little better than watching paint dry).

When it's done downloading, the iPhone SDK window appears onscreen, complete with an installer and various packages tied to the install process. All you then have to do is double-click the iPhone SDK installer and follow the (really simple) installation instructions. After you do all that, you'll have your very own iPhone Software Development Kit on your hard drive.

You'll become intimately acquainted with the iPhone SDK during the course of your project, but for now there's still one more bit of housekeeping to take care of: joining the official iPhone Developer Program. Read on to see how that works.

Joining the iPhone Developer Program

Okay, the simulator that comes standard with the iPhone SDK is a great tool for finding out how to program, but it does have some limitations. It doesn't support some hardware-dependent features, and when it comes to testing, it can't really emulate such everyday iPhone realities as CPU speed or memory throughput.

Minor annoyances, you might say, and you might be right. The real issue, however, is that just *registering* as a developer doesn't get you one very important thing — the ability to actually run your application on your iPhone, much less to distribute your application through Apple's iPhone App Store. (Remember that the App Store is the only way for commercial developers to distribute their applications to more than a few people.) To run your app on a real iPhone or get a chance to profile your app in the iPhone App store, you have to enroll in either the Standard or Enterprise version of the iPhone Developer Program. There's much speculation behind the reason for this, but the bottom line is that that's simply the way it is. At least (I might note) it isn't all that expensive.

It used to be that that the approval process could take a while, and although the process does seem quicker these days, it's still true that you can't run your applications on your iPhone until you're approved. You should enroll as early as possible.

If you go back to the iPhone Dev Center page, you can see a section in the right column that says iPhone Developer Program. (Refer to Figure 3-3.)

Figure 3-4 shows the page you're sent to in order to get some general information about the program. Use the Learn More links to explore as much as you'd like, but when you are ready, come back to this Overview page and click the Enroll Now button.

That button takes you to another page that asks you about your registered Apple Developer and Apple ID status. After you affirm that, yes, you are indeed a Registered Apple Developer, click Continue.

You're then taken to a page where you have to decide how you want to join the program — as an Individual or Company

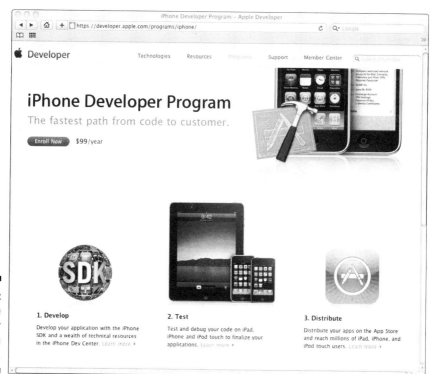

Figure 3-4:
The iPhone Developer program overview.

While joining as an individual is easier than joining as a company, there are clearly some advantages to enrolling as a company — for example, you can add team members (which I discuss in connection with the developer portal in Chapter 12), and your company name will appear in your listing in the App Store.

When you join as an individual, your real name shows up when the user buys (or downloads for free) your application in the App Store. If you're concerned about privacy, or if you want to seem "bigger," the extra work invoked in signing up as a company may be worthwhile for you.

I leave it up to you to decide which program you want. Personally, I started out as an Individual and then later created a company account as well.

Continue through the process, and eventually you'll be accepted in the developer program of your choice.

The next time you log in to the iPhone Dev Center, you'll notice that the page has changed somewhat. As a freshly minted Official iPhone Developer, you see the page shown in Figure 3-5. You find a number of iPhone Developer Program links on the right side, under iPhone Developer Program.

Here you can find links to the iPhone Provisioning Portal, which I cover in Chapter 12. I wouldn't go there now because it can be very confusing unless you understand the process. There's also a link for iTunes Connect, which you will use to upload your app and its information. I cover that in Chapter 12 as well. Underneath that, you can find links to Apple Developer Forums (more on that in a second) and the Developer Support Center.

Underneath that is a section titled App Store Resource Center. There is a wealth of information there, but I'd hold off looking through that for the time being for two reasons.

✔ **The app is the thing.** Until you actually develop an app, wading through the App Store policies, procedures, and resources is premature.

✔ **The amount of information can be overwhelming.** In Chapter 12, I give you an overview of how the whole process works. After that, you'll understand why all those resources are there and how to use them.

On the left side are two new links. There's information about the iAd JS Reference Library — Apple's new advertising network — and a link to the Apple Developer Forums. I'd be the first to say that developer forums can be very helpful, but I'd also be the first to admit that they are a great way to avoid doing other things, like working. As you scroll through the questions people have, be careful about some of the answers you see. No one is validating the information people are giving out. But take heart: Pretty soon you'll be able to answer some of those questions better yourself.

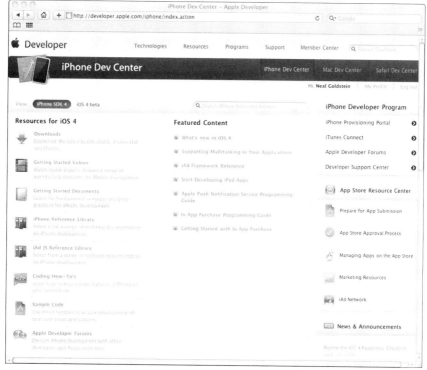

Figure 3-5:
Now you,
too, are
special —
the iPhone
Developer
Program
becomes
available.

Getting Yourself Ready for the SDK

Don't despair. I know the process is tedious, but it's over now. Going through this was definitely the *second* most annoying part of my journey toward developing software for the iPhone. The most annoying part was figuring out what Apple calls "provisioning" your iPhone — the hoops you have to jump through to actually run your application on a real, tangible, existing iPhone. I take you through the provisioning process in Chapter 12, and frankly, getting *that* process explained is worth the price of the book.

In the next chapter, I get you started using the SDK you just downloaded. I'm going to assume that you have some programming knowledge and that you also have some acquaintance with object-oriented programming, with some variant of C, such as C++, C#, and maybe even with Objective-C. If those assumptions miss the mark, help me out, okay? Take another look at the "Resources on the iPhone Dev Center" section, earlier in this chapter, for an overview of some of the resources that could help you get up to speed on some programming basics. Or, better yet, get yourself a copy of *Objective-C For Dummies.*

I'm also going to assume that you're familiar with the iPhone itself, and that you've explored Apple's included applications to become familiar with the iPhone's look and feel.

Part II
Using the iPhone Development Tools

The 5th Wave   By Rich Tennant

DISGUISED RINGTONES FOR NON-CELL PHONE VENUES

BROADWAY PLAY
Cough...cough...

CHURCH SERMON
ZZZZZ...
ZZZZZZ...

LIBRARY ©RICHTENNANT
SHHHH...
SHHHH...
SHHHH...

DENTIST'S WAITING ROOM
OW...OW...
OW...OW...
OW...OW...
OW...OW

In this part . . .

*W*hen you've established yourself as one of the developer in-crowd, you can download the SDK. Of course once you do that, you'll have to figure out how to use it, and you will. This part shows you how to download and use the iPhone Software Development Kit (SDK) and how to use Interface Builder — much more than your run-of-the-mill program for building graphical user interfaces — to start building a real interface. You can work along with me and then take all that knowledge and start working on your own app. Of course, before you get into the guts of coding your app, you need to know about what goes on during runtime inside those itty-bitty chips — and I take you through that as well.

Chapter 4

Getting to Know the SDK

. . . .

In This Chapter

▶ Getting a handle on the Xcode project

▶ Compiling an application

▶ Peeking inside the Simulator

▶ Checking out the Interface Builder

▶ Demystifying nib files

. . . .

I've said it before and I'll say it again: One of the things that really got me excited about the iPhone was how easy it was to develop applications. The Software Development Kit (SDK) comes with so many tools, you'd think developing must be really easy. Well, to be truthful, it's *relatively* easy.

In this chapter, I introduce you to the SDK. It's going to be a low-key, get-acquainted kind of affair. I show you the real nuts-and-bolts stuff in later chapters, when I actually develop the two sample applications.

Developing Using the SDK

The Software Development Kit (SDK) supports the kind of development process that's close to my heart: You can develop your applications without tying your brain up in knots. The development environment allows you to rapidly get a user interface up and running to see what it looks like. The idea here is to add your code incrementally — step by step — so you can always step back and see how what you just did affected the Big Picture. Your general steps in development would look something like this:

1. Start with Xcode, Apple's development environment for the OS X operating system.

2. Design the user interface.

3. Write the code.

4. Build and run your application.

5. Test your application.

6. Measure and tune your application's performance.

7. Do it all again until you are done.

In this chapter, I start at the very beginning, with the very first step, with Xcode. (Starting with Step 1? What a concept!) And the first step of the first step is to create a project.

Creating Your Project

To develop an iPhone application, you work in what's called an *Xcode project*. So, time to fire one up. Here's how it's done:

1. **Launch Xcode.**

   After you download the SDK, it's a snap to launch Xcode. By default, it's downloaded to /Developer/Applications, where you can track it down to launch it.

   Here are a couple of hints to make Xcode handier and more efficient:

   - If I were you, I'd drag the icon for the Xcode application all the way down to the Dock, so you can launch it from there. You'll be using it a lot, so it wouldn't hurt to be able to launch it from the Dock.

   - When you first launch Xcode, you see the Welcome screen shown in Figure 4-1. It's chock-full of links to the Apple Developer Connection and Xcode documentation. You may want to leave this screen up to make it easier to get to those links, but I usually close it. If you don't want to be bothered with the Welcome screen in the future, deselect the Show at Launch check box.

   Close the Welcome screen for now; you won't be using it.

2. **Choose File⇨New Project from the main menu to create a new project.**

   You can also just press Shift+⌘+N.

   No matter what you do to start a new project, you're greeted by the New Project window, as shown in Figure 4-2.

   The New Project window is where you get to choose the template you want for your new project. Note that the leftmost pane has three sections: one for the iPhone OS, one for Mac OS X, and a third for User Templates. You won't need to be working with user templates in this book (they allow you to add additional functionality to Xcode), so you can safely ignore it.

Figure 4-1:
The Xcode
Welcome
screen.

Figure 4-2:
The New
Project
window.

3. **In the New Project window, click Application under the iPhone OS heading.**

   The main pane of the New Project window refreshes, revealing several choices, as shown in Figure 4-2. Each of these choices is actually a template that, when chosen, generates some code to get you started.

4. **Select View-Based Application from the choices displayed and then click the Choose button.**

   Doing so brings up a standard Save sheet.

Note that when you select a template, a brief description of the template is displayed underneath the main pane. (Again, refer to Figure 4-2 to see a description of the View-Based Application. In fact, click some of the other template choices just to see how they're described as well. Just be sure to click the View-Based Application template again to get back to it when you're done exploring.)

5. **Enter a name for your new project in the Save As field, choose a Save location (the Desktop works just fine) and then click Save.**

I'm going to name my project ReturnMeTo. I suggest you do the same if you're following along with me.

After you click Save, Xcode creates the project and opens the Project window — which should look like what you see in Figure 4-3.

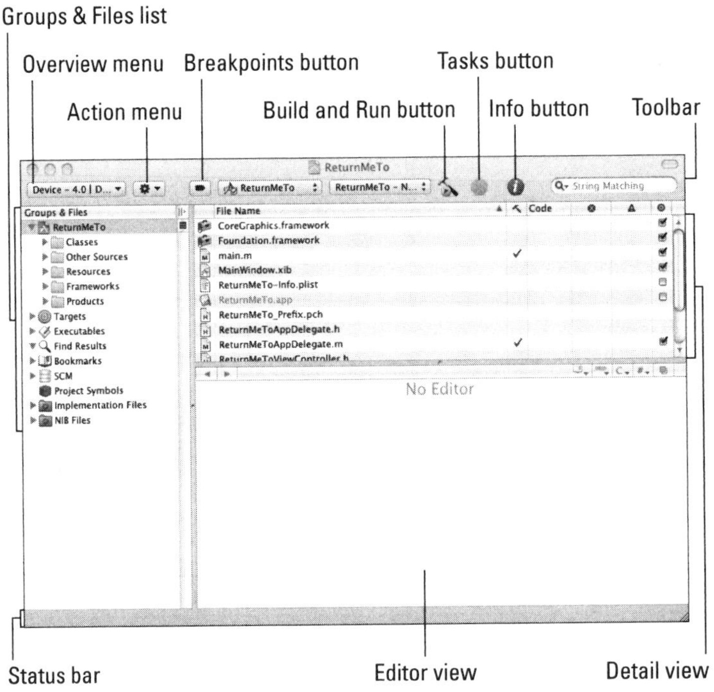

Groups & Files list
Overview menu   Breakpoints button   Tasks button
Action menu   Build and Run button   Info button   Toolbar

Status bar   Editor view   Detail view

Figure 4-3:
The ReturnMeTo Project window.

Exploring Your Project

To develop an iPhone application, you have to work within the context of an Xcode project. It turns out that you do most of your work on projects, using the Project window very much like the one in Figure 4-3. If you have a nice, large monitor, expand the Project window so you can see everything in it as big as life. This is, in effect, Command Central for developing your application; it displays and organizes your source files and the other resources needed to build your application.

If you take another peek at Figure 4-3, you see the following:

✔ **The Groups & Files list:** An outline view of everything in your project, containing all of your project's files — source code, frameworks, graphics, and some settings files. You can move files and folders around and add new folders. If you select an item in the Groups & Files list, the contents of the item are displayed in the topmost pane to the right — otherwise known as the Detail view.

Some of the items in the Groups & Files list are folders, whereas others are just icons. Most have a little triangle (the disclosure triangle) next to them. Clicking the little triangle to the left of a folder expands the folder to show what's in it. Click the triangle again to hide what it contains.

✔ **The Detail view:** Here you get detailed information about the item you selected in the Groups & Files list.

✔ **The Toolbar:** Here you can find quick access to the most common Xcode commands. You can customize the toolbar to your heart's content by right-clicking it and selecting Customize Toolbar from the contextual menu that appears. You can also choose View⇨Customize Toolbar.

   • Pressing the Build and Run button compiles, links, and launches your application.

   • The Breakpoints button turns breakpoints on and off and toggles the Build and Run button to Build and Debug. (I explain this in Chapter 10.)

   • The Tasks button allows you to stop the execution of your program that you've built.

   • The Info button opens a window that displays information and settings for your project.

✔ **The Status bar:** Look here for messages about your project. For example, when you're building your project, Xcode updates the status bar to show where you are in the process — and whether or not the process completed successfully.

✔ **The Favorites bar:** Works like other favorites bars you're certainly familiar with; so you can bookmark places in your project. This bar isn't displayed by default; to put it onscreen, choose View⇨Layout⇨Show Favorites Bar from the main menu.

✔ **The Text Editor navigation bar:** This navigation bar contains a number of shortcuts. These are shown in Figure 4-4. I explain more about them as you use them.

- *Bookmarks menu:* You create a bookmark by choosing Edit⟹Add to Bookmarks.

- *Breakpoints menu:* Lists the breakpoints in the current file — I cover breakpoints in Chapter 10.

- *Class Hierarchy menu:* The superclass of this class, the superclass of that superclass (if any), and so on.

- *Included Files menu:* Lists both the files included by the current file, as well as the files that include the current file.

- *Counterpart button:* This allows you to switch between header and implementation files.

✔ **The Editor view:** Displays a file you've selected, in either the Groups & Files list or Detail view. You can also edit your files here — after all, that's what you'd expect from the Editor view — although some folks prefer to double-click a file in the Groups & Files list or Detail view to open the file in a separate window.

To see how the Editor view works, check out Figure 4-5, where I've clicked the disclosure triangle next to the Classes folder in the Groups & Files list, and the `ReturnMeToAppDelegate.h` class in the Detail view. You can see the code for the class in the Editor view. (I deleted the comments you normally see when the template creates the classes and files for you.)

Also notice that I have customized my toolbar and added Active Target and Active Executable. I needed to do that for a project I am working on; you won't need to do that. But if you explore the options available to you to customize the toolbar you'll see you can set it up any way you like.

Clicking the Counterpart button switches you from the header (or interface) file to the implementation file, and vice versa. The header files define the class's interface by specifying the class declaration (and what it inherits from), instance variables (a variable defined in a class — at runtime all objects have their own copy), and methods. The implementation file, on the other hand, contains the code for each method.

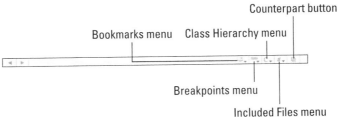

Figure 4-4: Text Editor navigation bar.

Counterpart button

Bookmarks menu   Class Hierarchy menu

Breakpoints menu

Included Files menu

Right under the Lock icon is another icon that lets you split the Editor view. That enables you to look at the interface and implementation files at the same time, or even the code for two different methods in the same or different classes.

If you have any questions about what something does, just position the mouse pointer above the icon, and a tooltip will explain it.

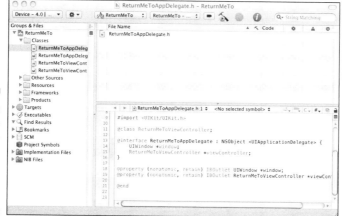

Figure 4-5:
The ReturnMe ToApp Delegate.h file in the Editor view.

The first item in the Groups & Files list, as you can see in Figure 4-6, is labeled ReturnMeTo. This is the container that contains all the source elements for my project, including source code, resource files, graphics, and a number of other pieces that will remain unmentioned for now (but I get into those in due course). You can see that your project container has five distinct groups (folders, if you will): Classes, Other Sources, Resources, Frameworks, Products. Here's what gets tossed into each group:

✔ **Classes** is where you should place all your code, although you are not obliged to. As you can see from Figure 4-5, this project has four distinct source-code files:

- ReturnMeToAppDelegate.h
- ReturnMeToAppDelegate.m
- ReturnMeToViewController.h
- ReturnMeToViewController.m

✔ **Other Sources** is where you'd typically find the precompiled headers of the frameworks you'll be using — stuff like ReturnMeTo_Prefix.pch as well as main.m, your application's main function.

✔ **Resources** contains, well, resources, such as `.xib` files, property lists images and other media files, and even some data files.

Whenever you choose the View-Based Application template (see Figure 4-2), Xcode creates the following three files for you:

- `ReturnMeToViewController.xib`
- `MainWindow.xib`
- `ReturnMeTo-Info.plist`

I explain `.xib` files in excruciating detail in this and following chapters. Soon you'll grow to love the `.xib` files as I much as I do.

✔ **Frameworks** are code libraries that act a lot like prefab building blocks for your code edifice. (I talk lots about frameworks in Chapter 2.) By choosing the View-Based Application template, you let Xcode know that it should add the `UIKit framework`, `Foundation.framework`, and `CoreGraphics.framework` to your project because it expects that you'll need them in a view-based application.

I'm going to limit myself to just these three frameworks in developing the ReturnMeTo Application. But I show you how to add a framework in Chapter 16.

✔ **Products** is a bit different from the previous three items in this list: It's not a source for your application, but rather *the compiled application itself.* In it, you find `ReturnMeTo.app`. At the moment, this file is listed in red because the file cannot be found (which makes sense, because you haven't built the application yet).

A file's name in red lets you know that Xcode can't find the underlying physical file.

If you happen to open the `ReturnMeTo` folder on your Mac, you won't see the folders that appear in the Xcode window. That's because those folders are simply logical groupings that help organize and find what you're looking for; this list of files can grow to be pretty large, even in a moderate-sized project.

After you build a lot of files, you'll have better luck finding things if you create subgroups within the Classes group and/or Resources group, or even whole new groups. You create subgroups (or even new groups) in the Groups & Files list by choosing New Project➪New Group from the main menu. You then can select a file and drag it to a new group or subgroup.

Building and Running Your Application

It's really a blast to see what you get when you build and run a project that you yourself created, using a template from the project creation window. Doing that is relatively simple:

1. **Choose Simulator | Debug from the Overview drop-down menu in the top-left corner of the Project window to set the active SDK and Active Build Configuration.**

   It may already be chosen for you. Here's what that means:

   - When you download an SDK, you actually download *multiple* SDKs — a Simulator SDK and a device SDK for each of the current iPhone OS releases.

   - Fortunately, for this book, I mostly use the Simulator SDK and iOS 4.0. Even more fortunately, in Chapter 12, I show you how to switch to the device SDK and download your application to a real-world iPhone. But before you do that, there's just one catch.

   - You have to be in the iPhone Developer Program to run your application on a device, even on your very own iPhone.

   A *build configuration* tells Xcode the purpose of the built product. You can choose between Debug, which has features to help with debugging (duh) and Release, which results in smaller and faster binaries. I use Debug for most of this book, so I recommend you go with Debug for now.

2. **Choose Build⇨Build and Run – Breakpoints On from the main menu to build and run the application. (A couple other options are available, as you can see, but this is the one I want you to work with).**

   You can also press ⌘+Y. This shortcut turns the green Build and Run button in the toolbar to orange, as it is in Figure 4-6 (which of course you can't see because the book is in black and white, but take my word for it). You won't be using breakpoints until Chapter 10, but you will frequently after that, and it's what you'll see in my screen shots. So you should do it now. You can also turn it back to green by clicking the Breakpoints button.

   Of course at this point you may have no idea what breakpoints are, but rest assured, I cover that all in Chapter 10,

   The status bar in the Project window tells you all about build progress, build errors such as compiler errors, or warnings — and (oh, yeah) whether the build was successful. Figure 4-6 shows that this was a successful build — you can tell by the tiny hammer icon and Succeeded in the *notification section* (on the right side) of the Status bar.

   Because you selected Debug for the active build configuration, the Debugger Console may launch for you, as shown in Figure 4-7 (I talk more about debugging in Chapter 10), depending on your Xcode preferences. (I get to them in a second.) If you don't see the console, select Run⇨Console to display it.

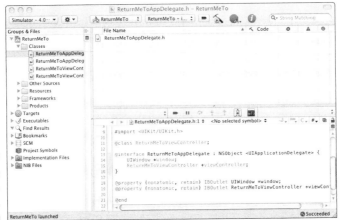

Figure 4-6: A successful build.

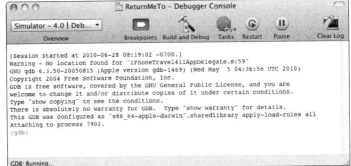

Figure 4-7: The Debugger Console.

But what if you make a mistake? If you enter the code *exactly* as I show you, there won't be any errors. But if you don't, I want to point out three kinds of things you might see. As I said, I won't be explaining the whys and wherefores of this until Chapter 10, but be aware of the following:

- If the compiler can't compile something, it lets you know by displaying the red exclamation point icon in the gutter to the left of the statement. (You can see a shining example in the top window of Figure 4-8.) If you click the red exclamation point icon, you get the compiler's justification. Read what the compiler tells you and go fix it.

- Sometimes the compiler can continue, but it doesn't like something about your code. In that case, it gives a warning (a yellow exclamation point icon, to be precise), as you can see in the bottom window in Figure 4-8. Again, click the warning icon to find out the compiler's concern.

- Your application may compile, and it may run into a problem during execution. As you see, that happened in the bottom window in Figure 4-8. (Warnings have a nasty habit of turning into execution errors, as you can see.)

Figure 4-8:
Warnings,
warnings
everywhere.

The red arrow you see next to the warning in Figure 4-8 (on top of the line number) points out the offending line of code (You can also see something called the Debugger strip — but ignore that for now). You get this red arrow because you compiled with Build and Run – Breakpoints On.

As I said, I don't explain these errors now, because if you enter the code exactly as shown, you won't be seeing any of these messages. But if you do make a slip and, as a consequence, see any variation of what you see in Figure 4-8, all you have to do is go back and enter the code exactly as shown, and all will be right with the world.

After it's launched in the Simulator, your first application looks a lot like what you see in Figure 4-9. You should see the status bar and a gray window, but that's it. (I know . . . this may look even more insipid than "Hello World," but I fix that big-time in Chapter 5.) You can also see the Hardware menu; I explain that next.

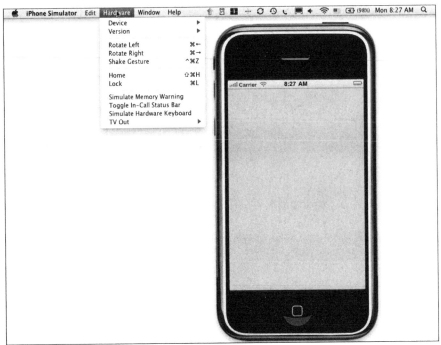

Figure 4-9:
Your first
application.

The iPhone Simulator

When you run your application, Xcode installs it on the iPhone Simulator
(or a real iPhone device if you specified the device as the active SDK) and
launches it. Using the Hardware menu and your keyboard and mouse, the
Simulator mimics most of what a user can do on a real iPhone, albeit with
some limitations that I point out shortly.

Hardware interaction

You use the iPhone Simulator Hardware menu (refer to Figure 4-8) when you
want your device to do the following:

- ✔ **Rotate Left:** Choosing Hardware⇨Rotate Left rotates the Simulator to
  the left. This enables you to see the Simulator in Landscape mode.

- ✔ **Rotate Right:** Choosing Hardware⇨Rotate Right rotates the Simulator to
  the right.

✔ **Use a Shake Gesture:** Choosing Hardware➪Shake Gesture simulates shaking the iPhone.

✔ **Go to the Home screen:** Choosing Hardware➪Home does the expected — you go to the home screen.

✔ **Lock the Simulator (device):** Choosing Hardware➪Lock locks the simulator.

✔ **Send the running application low-memory warnings:** Choosing Hardware➪Simulate Memory Warning fakes out your simulator by sending it a (fake) low-memory warning. I don't cover this, but it is a great feature for seeing how your app may function out there in the real world.

✔ **Toggle the status bar between its Normal state and its In Call state:** Choose Hardware➪Toggle In-Call Status Bar to check out how your application functions when the iPhone is not answering a call (Normal state) and when it supposedly *is* answering a call (In Call state).

The status bar becomes taller when you're on a call than when you're not. Choosing In Call state here shows you how things look when your application is launched while the user is on the phone.

✔ **Simulate Hardware Keyboard:** Choose Hardware➪Simulate Hardware Keyboard to simulate using a keyboard dock or wireless keyboard with an iPad device.

✔ **TV Out:** Choose Hardware➪TV Out *resolution* to bring up another window to simulate TV out (which allows you to mimic an external display) from the device.

The Hardware menu also allows you to specify the device (iPhone, iPad, iPhone 4) as well as the iOS version you want. I won't have you change those settings in this book.

Gestures

On the real device, a gesture is an action you do with your fingers to make something happen in the device, like a tap, or a drag, and so on. Table 4-1 shows you how to simulate gestures by using your mouse and keyboard.

Table 4-1	Gestures in the Simulator
Gesture	*iPhone Action*
Tap	Click the mouse.
Touch and hold	Hold down the mouse button.
Double tap	Double-click the mouse.
Two-finger tap	1. Move the mouse pointer over the place where you want to start.
	2. Hold down the Option key, which makes two circles appear that stand in for your fingers.
	3. Press the mouse button.
Swipe	1. Click where you want to start and hold the mouse button down.
	2. Move the mouse in the direction of the swipe and then release the mouse button.
Flick	1. Click where you want to start and hold the mouse button down.
	2. Move the mouse quickly in the direction of the flick and then release the mouse button.
Drag	1. Click where you want to start and hold the mouse button down.
	2. Move the mouse in the drag direction.
Pinch	1. Move the mouse pointer over the place where you want to start.
	2. Hold down the Option key, which will make two circles appear that stand in for your fingers.
	3. Hold down the mouse button and move the circles in or out.

Uninstalling applications and resetting your device

You uninstall applications on the Simulator the same way you do it on the iPhone, except you use the mouse rather than your finger.

1. **On the Home screen, place the pointer over the icon of the application you want to uninstall and hold down the mouse button until the icon starts to wiggle.**

2. **Click the icon's Close button — the little x that appears in the upper-left corner of the application's icon.**

3. **Click the Home button — the one with a little square in it, centered below the screen — to stop the icon's wiggling.**

You can also move an application icon around by clicking and dragging with the mouse.

You can remove an application from the background the same way you do it on the iPhone, except you use the mouse rather than your finger.

1. **Double-click the Home button to display the applications running in the background.**

2. **Place the pointer over the icon of the application you want to remove and hold down the mouse button until the icon starts to wiggle.**

3. **Click the icon's Remove button — the red circle with the – that appears in the upper-left corner of the application's icon.**

4. **Click the Home button to stop the icon's wiggling and then once again to return to the Home screen.**

To reset the Simulator to the original factory settings — which also removes all the applications you've installed — choose iPhone Simulator⇨Reset Content and Settings.

Limitations

Keep in mind that running applications in the iPhone Simulator is not the same thing as running them in the iPhone. Here's why:

✔ The Simulator uses Mac OS X versions of the low-level system frameworks, instead of the actual frameworks that run on the device.

✔ The Simulator uses the Mac hardware and memory. To really determine how your application is going to perform on an honest-to-goodness iPhone device, you're going to have to run it on a real iPhone device. (Lucky for you, I show you how to do that in Chapter 12.)

✔ Xcode installs applications in the iPhone Simulator automatically when you build your application by using the iPhone Simulator SDK (you saw that in Figure 4-9, for example). All fine and dandy, but there's no way to get Xcode to install applications from the App Store in the iPhone Simulator.

✔ You can't fake the iPhone Simulator into thinking it's lying on the beach at Waikiki. The location reported by the `CoreLocation` framework in the Simulator is fixed at

- • Latitude: 37.3317 North

- • Longitude: 122.0307 West

Which just so happens to be 1 Infinite Loop, Cupertino, CA 95014, and guess who "lives" there?

✔ **Maximum of two fingers.** If your application's user interface can respond to touch events involving more than two fingers, you need to test that on an actual device.

✔ You can access your computer's accelerometer (if it has one) through the UIKit framework. Its reading, however, will differ from the accelerometer readings on an iPhone (for some technical reasons I don't get into).

✔ OpenGL ES uses renderers on actual devices that are slightly different from those it uses on the iPhone Simulator. As a result, a scene on the Simulator and the same scene on a device may not be identical at the pixel level.

Customizing Xcode to Your Liking

Xcode offers options galore; many won't make any sense until you have quite a bit of programming experience, but a few are worth thinking about now.

1. **With Xcode open, choose Xcode⇨Preferences from the main menu.**

2. **Click Debugging in the toolbar, as shown in Figure 4-10.**

   The Xcode Preferences window refreshes to show the various preferences.

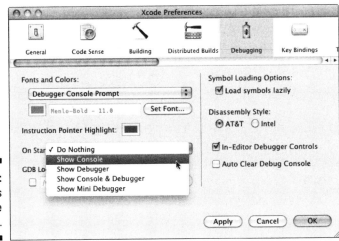

Figure 4-10:
Always show the console.

3. **Select the On Start drop-down menu and choose Show Console, as I've done in Figure 4-10. Then click Apply.**

   This automatically opens the Debugger Console after you build your application. This means that you won't have to take that extra step of opening the console to see your application's output.

4. **Click Building in the toolbar, as shown in Figure 4-11.**

5. **From the Build Results Window: Open during builds drop-down menu, choose Always, as I've done in Figure 4-11. Then click Apply.**

   This opens the Build Results window and keeps it open. You might not like this, but some people find it is easier to find and fix errors.

6. **Click the Documentation icon at the top, as shown in Figure 4-12.**

7. **Select the Check for and Install Updates Automatically check box and then press Check and Install Now.**

   This ensures that the documentation remains up-to-date. (It also allows you to load and access other documentation.)

8. **Click OK to close the Xcode Preferences window.**

You can also set the tab width and other formatting options by using the Indentation tab. I've set mine to 2 so I can display more on a page. The default is 4.

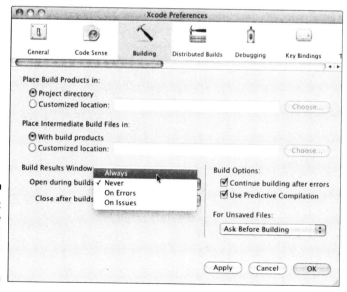

Figure 4-11:
Show
the Build
Results
window.

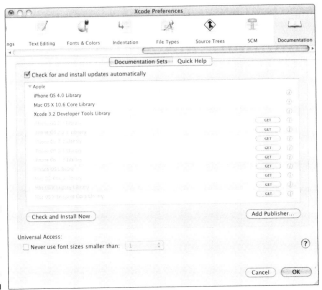

Figure 4-12:
Accessing
the docu-
mentation.

You can even have the Editor show line numbers. If you select Text Editing in the Xcode Preferences toolbar, you can select Show Line Numbers under Display Options. As you'll see, I leave them on all the time. The line numbers become very useful in the last part of Chapter 10.

Using Interface Builder

Interface Builder is a great tool for graphically laying out your user interface. You can use it to design your application's user interface and then save what you've done as a resource file, which is then loaded into your application at runtime. Then this resource file is used to automatically create the single window, as well as all your views and controls, and some of your application's other objects — view controllers, for example. (For more on view controllers and other application objects, check out Chapter 2.)

If you don't want to use Interface Builder, you can also create your objects programmatically — creating views and view controllers and even things like buttons and labels in your own application code. I show you how to do that as well. Often Interface Builder makes things easier, but sometimes just coding it is the best way.

Here's how Interface Builder works:

1. **In your Project window's Groups & Files list, expand the Resources group.**

2. **Double-click the `ReturnMeToViewController.xib` file, as shown in Figure 4-13.**

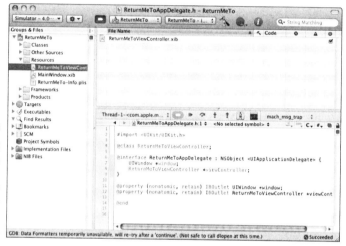

Figure 4-13: Selecting ReturnMe ToView Controller .xib.

Don't make the mistake of opening the `mainWindow.xib`. You need the `ReturnMeToViewController.xib` file.

Note that `ReturnMeToAppDelegate` is still in the Editor window; that's okay, because you're set to edit the `ReturnMeToViewController.xib` file in the Interface Builder, not in the Editor window. That's because double-clicking always opens a file in a new window — this time, the Interface Builder window.

What you see after double-clicking are the windows as they were the last time you left them. If this is the first time you've opened Interface Builder, you see three windows that look something like those in Figure 4-14.

Not surprisingly, the View window looks exactly as it did in the iPhone Simulator window — as blank as a whiteboard wiped clean.

Interface Builder supports two file types: an older format that uses the extension `.nib` and a newer format that uses the extension `.xib`. The iPhone project templates all use `.xib` files. Although the file extension is `.xib`, everyone still calls them *nib files*. The term *nib* and the corresponding file extension `.xib` are acronyms for NeXT Interface Builder. The Interface Builder application was originally developed at NeXT Computer, whose OPENSTEP operating system was used as the basis for creating Mac OS X.

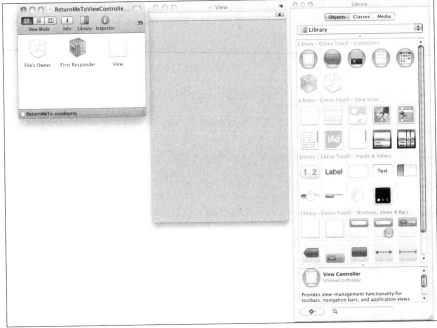

Figure 4-14:
The
ReturnMe
ToView
Controller
in Interface
Builder.

The window labeled `ReturnMeToViewController.xib` (the far-left window in Figure 4-14) is the nib's main window. It acts as a table of contents for the nib file. With the exception of the first two icons (File's Owner and First Responder), every icon in this window (in this case, there's only one, but you'll find more as you get into nib files) represents a single instance of an Objective-C class that will be created automatically for you when this nib file is loaded.

Interface Builder does not generate any code that you have to modify or even look at. Instead, it creates instant Objective-C objects that the nib loading code reconstitutes and turns into real objects at runtime.

If you were to take a closer look at the three objects in the `ReturnMeToView Controller.xib` file window — and if you had a pal who knew the iPhone backwards and forwards — you'd find out the following about each object:

✔ **The File's Owner proxy object:** This is the controller object that's responsible for the contents of the nib file. In this case, the File's Owner object is actually the `ReturnMeToViewController` that was created by Xcode and will be the primary object you'll use to implement the application's functionality. The File's Owner isn't created from the nib file. It's created in one of two ways: either from another (previous) nib file or by a programmer who codes it manually.

In Interface Builder, you can create connections between the File's Owner and the other interface objects in your nib file. For example, in Chapter 7, I create a connection between the `ReturnMeToView Controller` and a text field (for entering a phone number) and a label (in which to display the phone number).

✔ **First Responder proxy object:** This object is the first entry in an application's dynamically constructed responder chain (a term I explain in Chapter 6) and is the object with which the user is currently interacting. For a view, it's usually going to start out as the View Controller object. If, for example, the user taps a text field to enter some data, the first responder would then become the Text Field object.

Although I use the first responder mechanism quite a bit as I build the ReturnMeTo application, there's actually nothing I have to do to manage it. It's automatically set and maintained by the `UIKit` framework.

✔ **View object:** The View icon represents an instance of the `UIView` class. A `UIView` object is an area that a user can see and interact with. In this application, you'll only have to deal with the one view.

If you take another look at Figure 4-14, you'll notice two other windows open besides the main window. Look at the window that has the word View in the title bar. That window is the graphical representation of the View icon. If you close the View window and then double-click the View icon, this window will open up again. This is your canvas for creating your user interface: It's where you drag user-interface elements such as buttons and text fields. These objects come from the Library window (the third window you see in Figure 4-14).

The Library window contains your palette — the stock Cocoa Touch objects that Interface Builder supports. Dragging an item from the Library to the View window adds an object of that type to the View (and, remember, it's added as a subview).

If you happen to close the Library window, whether by accident or by design, you can get it to reappear by choosing Tools⇨Library.

It's Time to Get to Work

Finally, at long last (actually, it's been only a few pages), you're ready to do some real work. In the next chapter, I lead you through creating the user interface for the ReturnMeTo application.

So take a break if you need to, but come back ready to work.

Chapter 5

Building the User Interface

*A*s I've mentioned before and will say many times again (unless the editors stop me), the user interface, although critical for most applications, is less forgiving on the iPhone than on the desktop. That's because onscreen real estate is limited on the device. (Come on, as cool as the screen is, it's still smaller than a desktop monitor.) Given the space limitations, I always like to get a pretty good idea of what the user interface will be like, because it could have a definite impact on my software architecture. Before I start coding, I want to be sure that the interface is going to work in its intended space.

When I started the ReturnMeTo application, I thought the user interface would be a piece of cake — and to some extent, it was. But even the easiest of applications — apps like ReturnMeTo that have focused functionality and a single window — can benefit from a little road-testing. As I go through the process of developing this application over the next few chapters (and take it for a spin now and then), I document what happens along the way — the good, the bad, and the ugly. There's method to this madness: As I try different implementations of the application, you get a close look at how easy it is to make those changes.

Starting Interface Builder

First things first: Start up Interface Builder so you can start laying out the user interface. Just so you know what you're aiming for, Figure 5-1 shows what the final application is going to look like in the Simulator.

Isn't it a beauty? All modesty aside, you, too, can build cool-looking apps in no time. Here's what you need to do:

1. **Launch Xcode.**

   It's located in /Developer/Applications. (If you listened to my advice in Chapter 3 and added the Xcode icon to the Dock, you can of course launch it from there.)

2. **With Xcode open, choose File⇨Open from the main menu and then use the Open dialog to navigate to and open the ReturnMeTo project created in Chapter 4.**

   The ReturnMeTo Project window appears onscreen.

   If you haven't created the project yet, check out Chapter 4 — or, after you've launched Xcode, follow these steps:

   a. Choose File⇨New Project.

   You're asked to choose a template for your project.

b. Choose a View-Based application.

c. Name the application **ReturnMeTo** and then save it.

3. **In the Groups & Files list (on the left side of the Project window), click the triangle next to the Resources folder to expand it, as shown in Figure 5-2.**

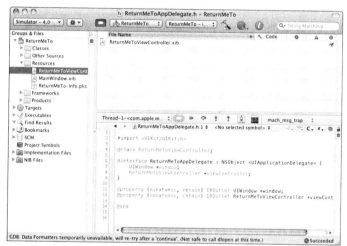

Figure 5-2:
The Project
window.

4. **In the expanded Resources folder, double-click the `ReturnMeToViewController.xib` file.**

   Doing so launches Interface Builder — and if you've never run this program before, you end up with something that looks like Figure 5-3. (If you already spent some time exploring Interface Builder, you see the windows as you last left them.)

5. **Check to see whether the Library window (at the right in Figure 5-3) is open. If it isn't, open it by choosing Tools⇨Library or ⌘+Shift+L. Make sure that Objects is selected in the mode selector at the top of the Library window and Library in the drop-down menu right below the mode selector.**

   The Library has all the components you can use to build a user interface. These include the things you see on the iPhone screen, such as labels, buttons, and text fields; and those you need to create the "plumbing" to support the views (and your model) such as the view controller I explain in Chapter 2.

   `ReturnMeToViewController.xib` was created by Xcode when I created the project from the template. As you can see, the file already contains a view — all I have to do here is add the static text, images, and text fields. If you drag one of these objects to the View window, it creates that object when your application is launched.

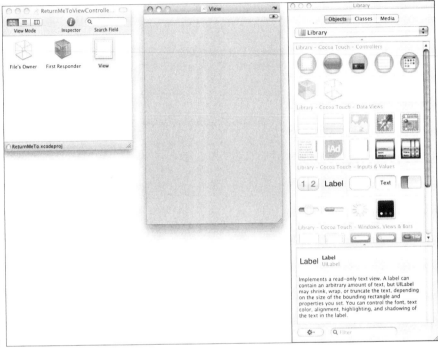

Figure 5-3:
Interface
Builder
windows.

6. **Drag the Label element from the Library window over to the View window, as shown in Figure 5-4.**

   Labels display static text in the view (*static text* can't be edited by the user).

   You may notice a rectangle around the label when you're done dragging it over to the View window. This rectangle won't show onscreen when the app is running. (You can turn this particular feature on or off, as I have, by choosing Layout➪Show/Hide Bounds Rectangle.)

   Your View should look something like Figure 5-4 when you're done.

   Labels are actually subviews of your main view. Knowing that will make it a lot easier to understand some things I end up doing in the next few chapters. I remind you of this whole view/subview thing when it comes up again.

7. **Click to select the Label text and then choose Tools➪Attributes Inspector.**

   The Attributes Inspector appears onscreen, as shown in Figure 5-5.

   Pressing ⌘+1 is another way to call up the Attributes Inspector.

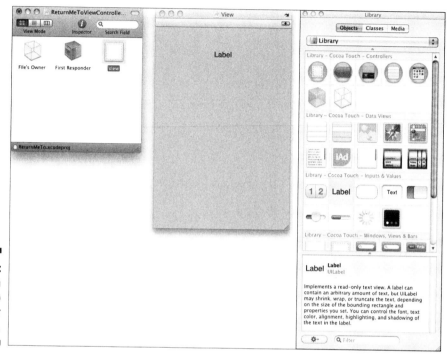

Figure 5-4:
Adding a
label to
the user
interface.

Note the four icons across the top of the Attributes Inspector window. They correspond to the Attributes, Connections, Size and Identity Inspectors, respectively, in the Tools menu. The Attributes icon looks pushed down in Figure 5-5, which makes sense because the Attributes Inspector is the active one.

8. Enter If you find me **in the Attributes Inspector's Text field, enter 34 in the Font Size field, and then press Return or Enter.**

The minimum font size business can be confusing. Notice in the Attribute Inspector that the Adjust to Fit check box is selected. That means entering a font size here is really saying, "I want the text to fit in the label, but don't make it any smaller than the size I've entered here in Font Size." The resulting font size is a side effect of setting the minimum larger than the label. The problem is that if you deselect the Adjust to Fit box and you want to specify the actual font size, using the Font Size field can increase it but not decrease it. (All you're doing in the Font Size field is setting the minimum, and it's already larger than that.) To actually specify a smaller font size, you have to choose Font⇨Show Fonts and set it from there.

Connections   Identity

Attributes   Size

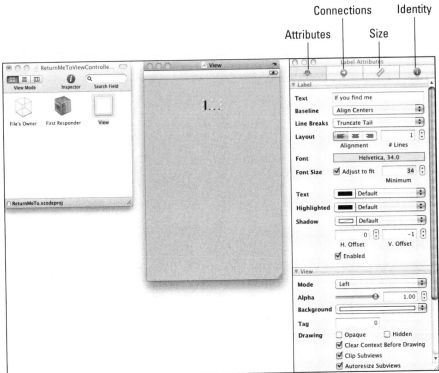

Figure 5-5:
Formatting
the label.

Okay, at this point I could double-click the label in the View window and enter the new text, but I need the Attributes Inspector to change the font size. So I might as well change the text in the Attributes Inspector. (I can also change other attributes there, such as color, if necessary.)

What you see in response is something like Figure 5-5. Not very appealing is it? Where the heck is the text, for example? It turns out that you have to increase the size of the label to see the larger text.

9. **With the label still selected, choose Layout⇨Size to Fit to increase the size of the label.**

I could select the label and resize it by dragging the selection points you see in Figure 5-5, but I'm lazy enough just to choose Layout⇨Size To Fit, which does exactly that: It adjusts the label size to fit the text.

10. **Choose Layout⇨Alignment⇨Align Horizontal Center in Container to center the label in the View screen.**

You have a couple of ways to center things in the View. If you had some other objects in place that were already centered, for example, you could use the guides provided by Interface Builder. (You can see them in Figure 5-10, if you peek ahead a few pages.) But in this case, there are no other objects to use as a reference; you're better off just using the Layout menu.

Time now to specify a more appealing color for your View screen's background.

11. **With the view itself (rather than just the label) selected, select the Background drop-down menu in the Attributes Inspector.**

   Note that selecting the view instead of the label changes the composition of the Attributes Inspector.

12. **Select Default from the menu (you see a white box next to it) to change the View background from gray to white, as shown in Figure 5-6.**

13. **Choose File⇨Save to save what you have done.**

   You can also save your work by pressing ⌘+S.

   Ready to admire your work? For that, you need to build and run your application in Xcode.

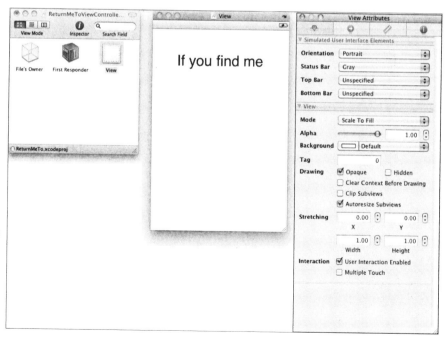

Figure 5-6:
Changing
the
background
color.

Be sure to save your work. Forgetting to save your work has caused many developers (including yours truly) to waste prodigious amounts of time trying to figure out why something "doesn't work."

14. Make your Xcode window the active window again.

If you can't find it, or you minimized it, just click the Xcode icon on the Dock. The ReturnMeTo project should still be the active one. (You can always tell the active project by looking at the project name at the top of the Groups & Files list.)

15. In Xcode, click the Build and Run button in the Project Window toolbar. (Refer to Figure 5-2.)

You can also choose Build⇒Build and Debug from the main menu or press ⌘+Return.

The Simulator launches automatically. It shows you something like Figure 5-7, depending on how creative you've been. (Maybe you went with Cranberry Red as your background color rather than the default or white?)

Figure 5-7:
Admiring
your work.

This is the general pattern I use as I build my interface — add stuff, build and run, and then check the results. Although I don't run the program in the Simulator after every change (unless, of course, I'm trying to avoid doing something else), I do run it periodically to check what the program will look like on the iPhone.

You can also choose File⇨Simulate Interface from the Interface Builder menu to see what the interface will look like.

Adding Graphics and the Rest of the Elements

All background color and text makes Jack a dull boy, so you'll definitely want to add some graphics to keep Jack's (or Jill's) interest. Originally, I created my own little iPhone graphic (I explain why I didn't use it in Chapter 12 — I'm going to use a royalty-free image I found on the Internet instead), and I show you how to get it safely into an application. (I admit that I didn't give up a promising career as a graphic artist to become a software developer, but it's the thought that counts, right?)

To get an image placed in your application, first you need (well, yeah) an image.

The preferred format for the image is .png. Although most common image formats display correctly, Xcode automatically optimizes .png images at build time to make them the fastest and most efficient image type for use in iPhone applications.

After you have your image, do the following:

1. **Back in Xcode, drag the graphics file into the Resources folder in the Project window, as shown in Figure 5-8.**

   Xcode asks you whether you want to make a copy of the icon file. Otherwise, it will simply create a pointer to the file. The advantage of using a pointer is that if you modify the image later, Xcode will use that modified version. The disadvantage is that Xcode won't be able to find the image file if you move it.

   I'm all for copying.

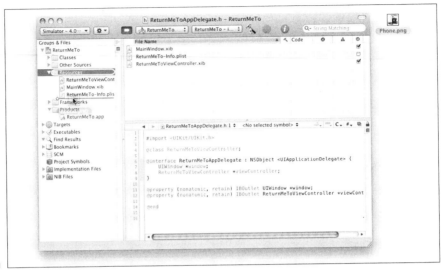

Figure 5-8:
Dragging a
file into the
Resources
folder.

2. **Select the Copy Items into Destination Group's Folder check box to copy the file, as shown in Figure 5-9.**

   An alternative is to click the Resources folder in Xcode, choose Project⇨Add to Project, and then navigate to the file you want to add.

Figure 5-9:
Copying the
image to the
Resources
folder.

3. **Return to Interface Builder and make sure that Objects is selected in the mode selector at the top of the Library window and that Library is selected in the drop-down menu right below the mode selector.**

You can manage this return by clicking in an Interface Builder window or by clicking the Interface Builder icon on the Dock.

4. Drag the Image View element from the Library onto the View window.

5. Select the Image View element in the View window.

Doing so changes what you see in the Attribute Inspector. It now displays the attributes for an Image view, as shown in Figure 5-10.

Notice the blue lines displayed by Interface Builder. They're there to make it easy to center the image. Interface Builder also displays blue lines at the borders to help you conform to Apple User Interface Guidelines. If you jump ahead to Figure 5-14, you can see the borderlines as plain as day.

6. Using the inspector's Image drop-down menu, choose the image you want to use for the image view, as shown in Figure 5-11.

That inspector's a handy little critter, isn't it?

If you don't see Phone.png — the image file you added way back in Step 1 — in the drop-down menu, select File⇨Reload All Class Files.

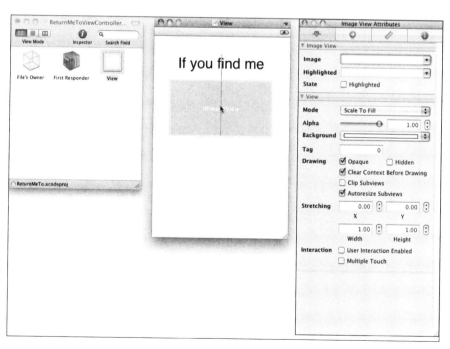

Figure 5-10:
Centering
the image.

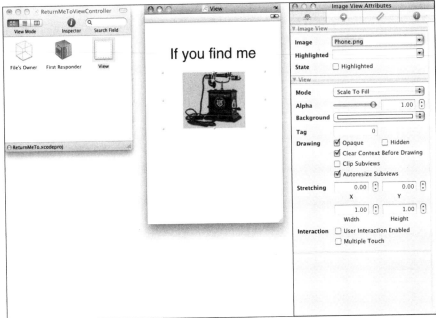

Figure 5-11:
Selecting
the image.

7. **Choose Layout➪Size to Fit and Layout➪Alignment➪Align Horizontal Center in Container to align and center the image.**

8. **Using the steps included in the previous section for adding the "If you find me" text, add the "Please call" text.**

You know the drill: Drag a label from the Library window, add the "Please call" text, increase the font size to 34, and then choose Layout➪Size to Fit and Layout➪Alignment➪Align Horizontal Center in Container. Click the Text field (not the drop-down menu) in the Attributes Inspector to call up the little crayon box you see in Figure 5-12, and click the red crayon to make the text red so it stands out more. If the crayons remind you too much of kindergarten, you can get other color-palette views by clicking the icons at the top of the Colors window. If you do click the drop-down menu instead, select Other, and that brings up the crayon box as well.

You can also see a red line across the view that the "Please call" text sits on. You can get that by choosing Layout➪Add Horizontal Guide (there's a vertical one as well). These guides can help you line things up.

Figure 5-12:
The red
"Please
call" label.

9. **Add yet another label to display the number to call.**

This next label needs to hold the text of the number the user enters. (As you see later, I also use it to communicate some information to the user, so it needs to be able to fit more text than just a phone number.) Just drag a label object from the Library into the view, but this time, because you don't know the size of the text you're going to display, widen it to almost the width of the view. To do that, use the selection points you see in Figure 5-13 to resize the label. (The blue lines shown in Figure 5-14 help you stay within the iPhone Human Interface Guidelines.)

You want the text centered in the view, but because this time the text may be smaller than the view, you should make sure that it's centered by clicking the Align Center control in the Layout field of the Attributes Inspector, as shown in Figure 5-13.

You could leave the text field blank, or put in some default text for the user to see before he or she has entered a phone number the first time. I've done it both ways and found (after some user testing) that people preferred that they see some text when they first launch the application, so I added "650 555 1212." No matter what you do, make sure that the Align Center control is chosen.

Align Center

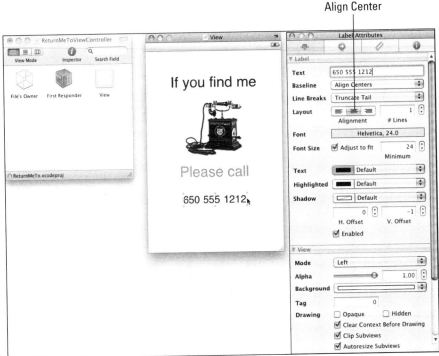

Figure 5-13:
Dragging
the label's
selection
points.

Label fields are not editable by the user, but you can change the text in your program.

10. **Drag a Text Field element from the Library (refer to Figure 5-4) into the View window to add a text-entry field, as shown in Figure 5-15.**

    This is the field a user will use to enter his or her phone number. Go ahead and put it under the Phone Number label.

11. **Using the text field's selection points, extend both sides of the field so that it ends up matching the field above.**

    If you click into the text field, you see both a Text and Placeholder field in the Attributes Inspector. The Text field specifies what the user will see before he or she touches the text field to enter something. The Placeholder, on the other hand, is a default text that the user will see after he or she touches the text field to enter text. Personally, I find the default annoying, so I left it blank. I also prefer text entry to be aligned left, which is the default for the Text field.

12. **Set the Keyboard Type to Numbers & Punctuation, as shown in Figure 5-15.**

Figure 5-14:
Formatting
the number
to call label.

Figure 5-15:
Adding the
text entry
field number
to call.

13. Make your Xcode window the active window again and then click the Build and Run icon in the Project Window toolbar.

You can also choose Build⇨Build and Run from the main menu or press ⌘+Return.

I bet you've been waiting for this moment with bated breath and can hardly wait to see what the application looks like. Your application should look like Figure 5-16, depending on which (if any) liberties you have taken.

Figure 5-16:
The completed user interface in the Simulator.

Adding an Application Icon

One of the design goals for the ReturnMeTo application was to make it obvious to someone who found my iPhone that he or she should touch the application icon.

But if you click the Simulator's home button — the black button with the white square at the very bottom of the window — what you see in Figure 5-17 is an application icon that's noticeable only for what it is not. What I need is an icon that reaches out and says, "touch me!"

Figure 5-17:
The
"stylish"
ReturnMeTo
icon.

An application icon is simply a 57-x-57-pixel .png file, just like the one I used for our image (albeit smaller) in the "Adding Graphics and the Rest of the Elements" section, earlier in the chapter. I created an icon matching those measurements in a graphics program and added it to the ReturnMeTo project in the same way I added the image file earlier — by dragging it into the Resources folder.

After I add the icon, I also need to specify that this icon is what I want used as the application's icon. I do that by using one of those other mysterious files you see in the Resources folder. Here's how:

1. **In the Resources folder, click the info.plist file, as shown in Figure 5-18.**

   The contents of the info.plist file are displayed in the Editor pane. You're treated to some information about the application, including an item in the Key column labeled Icon File.

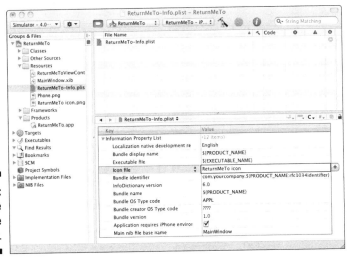

Figure 5-18: Adding the icon in the info.plist.

2. **Double-click in the empty space in the Value column next to Icon File.**

3. **Type in ReturnMeTo icon.png (or whatever name you chose to give your image) and then build the project as you normally would.**

   You know, clicking the Build and Run button in the Project Window toolbar, choosing Build⇨Build and Go (Run) from the main menu or pressing ⌘+Return.

I do more with the info.plist file and its various settings when I get ReturnMeTo ready for the App store in Chapter 12.

Click the Home button, and you should be able to see your application icon. (For a peek at mine, check out Figure 5-19.)

Figure 5-19:
The
ReturnMeTo
icon on the
iPhone.

A Lot Accomplished Very Quickly

In only a few pages, you've accomplished quite a bit. I want to emphasize, as you will see, that what I've done isn't just a design. What you see here is the specification for code that will take what I've laid out in Interface Builder and create the objects that will implement it at runtime.

I do have to write code, however, if I want the application to actually do something. But before I get into that subject in detail, it's important that you understand how the application works at runtime — how the objects fit together and communicate. I explain what you need to know about that whole business in the next chapter.

Chapter 6

While Your Application Is Running

. . . .

In This Chapter

▶ Seeing how applications actually work

▶ Getting a handle on how nib files work

▶ Following what goes on when the user taps your application icon

▶ Managing events

▶ Creating an app with a view

▶ Remembering memory management

▶ Knowing what else you should be aware of at runtime

. . . .

*T*aking a peek at the iPhone application architecture (Chapter 2) and working through the steps of creating the user interface for an application (Chapter 5) are all fine and dandy, but at some point, you have to add some code — and to do that, you need an additional frame of reference: You need to know how all this stuff works at runtime.

Uncovering the mysteries of runtime is the goal of this chapter.

Okay, if you can't wait to code, then by all means skip to the next chapter. Honestly, if I were you (or me without benefit of hindsight), I'd be itching to do the same thing. But you'll probably run into trouble, just as I did, if you don't take some time beforehand to examine how the objects work together to deliver the user's experience of the application.

Application Anatomy 101 — The Life Cycle

The short-but-happy life of an application begins when a user launches it by tapping its icon on the Home screen. The system launches your application by calling its `main` function.

The `main` function does only three things:

✔ Sets up an autorelease pool.

✔ Calls the `UIApplicationMain` function.

✔ At termination, releases the autorelease pool.

To be honest, this whole `main` function thing isn't something you even need to think about. What's important is what happens *after* the `UIApplicationMain` function is called. The whole ball of wax is shown in Figure 6-1. Here's the play-by-play:

1. **The main nib file is loaded.**

   A *nib file* is a resource file that contains the specifications for one or more objects. The main nib file usually contains a window object of some kind, the application delegate object, and any other key objects. When the file is loaded, the objects are reconstituted (think "instant application") in memory.

   For the ReturnMeTo example, this is the moment of truth when the `ReturnMeToAppDelegate`, `ReturnMeToViewController`, its view, and the main window are created.

   For more on those objects and the roles they play in applications, see Chapter 2.

2. **The `UIApplicationMain` sends the *application delegate* the `application:didFinishLaunchingWithOptions:` message.**

   This step is where you initialize and set up your application. You have a choice here: You may want to display your main application window as if the user was starting from scratch, or you may want the window to look the way it did when the user last exited the application. The application delegate object is a custom object that you code. It's responsible for some of the application-level behavior of your application. (Delegation is an extensively used design pattern that I explain in Chapter 2.)

3. **The `UIKit` framework sets up the event loop.**

   The *event loop* is the code responsible for polling input sources — the screen, for example. Events, such as touches on the screen, are sent to the object — say, a controller — that you have specified to handle that kind of event. These handling objects contain the code that implements what you want your application to do in response to that particular event. A touch on a control may result in a change in what the user sees in a view, a switch to a new view, or even the playing of "My Melancholy Baby."

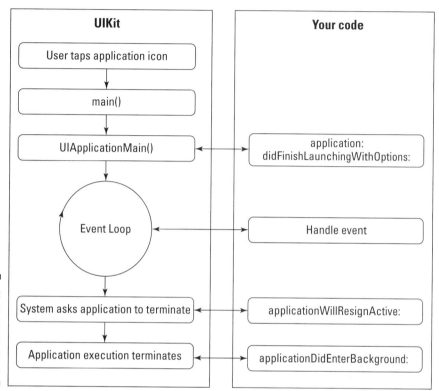

Figure 6-1:
A simplified
lifecycle
view of an
iPhone
application.

4. The normal processing of your application is interrupted.

In OS 4.0, your application must be able to handle those situations
where its normal processing is interrupted. The interruption may be
temporary — an incoming phone call or SMS message, for example — or
it may be a permanent one, such as when the user switches out of the
application and your application begins the transition to the background
state.

The OS sends you a number of messages to let you know exactly what's
happening and also to give you the opportunity to do things such as
save user data and *state information* — saving where the user was in
the application. (I cover all such messages in the "Normal processing is
interrupted" section, later in this chapter.)

Saving is important, because if the application is terminated, when it is
launched again (refer to Step 2) and the UIApplicationMain sends
the application delegate the applicationDidFinishLaunchingWith
Options message, you can restore the application to where the user
left off.

5. **Your application can be terminated.**

   Under OS 4.0, the normal course of events is for applications not to be terminated — they are simply moved to the background. But under some circumstances, your application can in fact be terminated. (I take the time to explain what those circumstances are in the "Application termination" section, later in this chapter.)

It all starts with the main nib file

When you create a new project by using a template — quite the normal state of affairs, as I explain in Chapter 3 — the basic application environment is already included. That means when you launch your application, an application object is created and connected to the window object, the run loop is established, and so on — despite the fact that you haven't done a lick of coding. Most of this work is done by the UIApplicationMain function as illustrated back in Figure 6-1.

But what does the UIApplicationMain function actually do? I'm glad you asked. When it goes through its paces, the process works more or less as follows, as illustrated in Figure 6-2.

1. **An instance of UIApplication is created.**

2. **UIApplication looks in the Info.plist file, trying to find the main nib file.**

   It makes its way down the Key column until it finds the Main Nib File Base Name entry. Eureka! It peeks over at the Value column and sees that the value for the Main Nib File Base Name entry is MainWindow.

3. **UIApplication loads MainWindow.xib.**

The file MainWindow.xib is what causes your application's delegate, main window, and view controller instances to get created at runtime. Remember, this file is provided as part of the project template. You don't need to change or do anything here. This is just a chance to see what's going on behind the scenes.

To take advantage of this once-in-a-lifetime opportunity, go back to your project window in Xcode, expand the Resources folder in the Groups & Files list on the left, and then double-click MainWindow.xib. (You do have a project, right? If not, check out Chapter 4.) When Interface Builder opens, take a look at the nib file's main window — the one labeled MainWindow.xib, which should look like the MainWindow.xib you see in Figure 6-2. Double-click the ReturnMeToViewController object as well as the Window object, if they aren't already open. You should end up with three windows open, as shown in Figure 6-3.

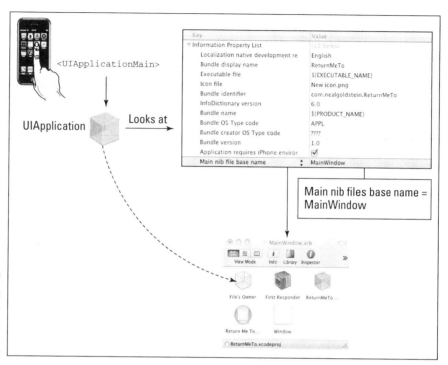

Figure 6-3 shows that `MainWindow.xib` contains five files. The objects you see are as follows:

- ✔ **A File's Owner proxy object:** The File's Owner object is actually the `UIApplication` instance. This object isn't created when the file is loaded as are the window and views. It's already created by the `UIApplicationMain` object before the nib file is loaded.

- ✔ **First Responder proxy object:** This object is the first entry in an application's responder chain, which is constantly updated while the application is running to (usually) point to the object with which the user is currently interacting. If, for example, the user were to tap a text field to enter some data, the first responder would become the text field object.

- ✔ **An instance of `ReturnMeToAppDelegate` set to be the application's delegate.**

- ✔ **An instance of the `ReturnMeToViewController`.**

- ✔ **A window:** The window has its background set to white and is set to be visible at launch. This is the window you see when the application launches.

Okay, so all these disparate parts of the `MainWindow.xib` are loaded by `UIApplication`. What happens next is shown in Figure 6-4. The numbers in the figure correspond to the following steps:

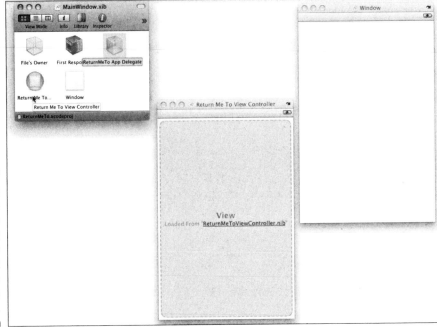

Figure 6-3:
The application's Main Window.xib as it appears in Interface Builder.

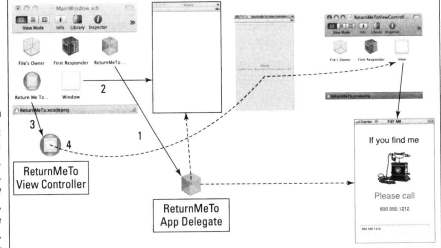

Figure 6-4:
Creating the AppDelegate, window, view controller, and the view.

1. Create `ReturnMeToAppDelegate`.

2. Create `Window`.

3. Create `ReturnMeToViewController`.

4. `ReturnMeToViewController:LoadView` loads the view from the `ReturnMeToViewController.xib` file.

Wait a sec — how does the `ReturnMeToViewController` know that it's supposed to do that? If you double-click the `ReturnMeToView Controller` object in the `MainWindow.xib` window (refer to Figure 6-3), you can see the `ReturnMeToViewController` window with its view. There, right in the middle of that view, it tells you that it will be "Loaded from `ReturnMeToViewController.nib`." (As I explain in Chapter 4, a nib file type used to be the term of choice. Here's a vestige of that. Don't worry; despite the nib business, it really will be the `.xib` file.) If you use the Inspector to look at the view controller attributes (see Figure 6-5), you can see that the NIB Name drop-down list specifies said NIB name — the nib file for the view controller. In this case, you see `ReturnMeToViewController` specified. This makes the connection explicit between the `MainWindow.xib` and the `ReturnMeToViewController.xib`. All the work is done without any fuss or bother, I might add, by Xcode when you created the project from the template.

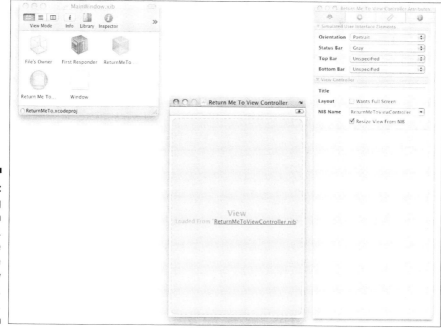

Figure 6-5:
Connecting
the Main
Window.
xib and the
ReturnMe
ToView
Controller.
xib.

Initialization

The next step is for the UIApplication to send the ReturnMeToAppDelegate applicationDidFinishLaunching: message. Step 5 in Figure 6-6 represents what happens when the application:didFinishLaunchingWithOptions: method is invoked.

This figure looks exactly like Figure 6-4, except I added a Step 5: putting the view into the window and then making the window visible. At this point, you'd do any other application initialization as well — *and* return everything to what it was like when the user last used the application.

When the application:didFinishLaunchingWithOptions: method is invoked, your application is in the inactive state. Unless your application does some kind of background processing, when your application becomes active, it will receive the applicationDidBecomeActive: message when it *enters the foreground* (becomes the application the user sees on the screen).

Under OS 4.0, an application can also be launched into the background, but because the ReturnMeTo application at the heart of this section doesn't do any background processing, this is the sequence I work with. And because the application does no background processing, there's also nothing it has to do in response to the applicationDidBecomeActive: message.

Feeling cheated by this lack of background processing? Not to worry; when I develop the iPhoneTravel411 application in later chapters, I show you how to develop an application that also does some processing in the background and I explain all the permutations and combinations of foreground and background messages that you need to take into account to run in both background and foreground.

Figure 6-6 is more than just a conceptual diagram, as you can see from Listings 6-1 and 6-2, which show two instance variables, window, and viewController. These are automatically filled in for you when your application is launched. Then in the application:didFinishLaunchingWithOptions: method, the view controller and its view are added to the window, and the window becomes visible. Note that this was generated automatically by Xcode. I get into what all those strange things like IBOutlet and @property mean in the next chapter, and I talk about some of the methods you see automatically generated for you here in a bit — actually, in the "Normal processing is interrupted" section, later in this chapter.

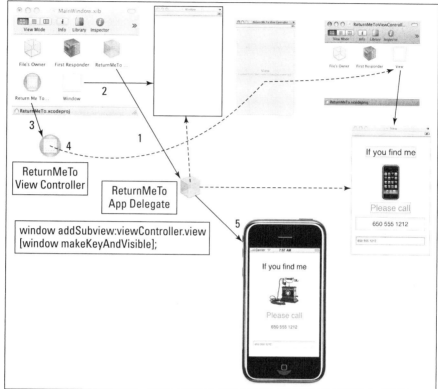

Figure 6-6:
The
application
DidFinish
Launching:
method
installs the
view in the
window and
makes the
window
visible.

Listing 6-1: ReturnMeToAppDelegate.h

```objc
#import <UIKit/UIKit.h>

@class ReturnMeToViewController;

@interface ReturnMeToAppDelegate : NSObject
        <UIApplicationDelegate> {
    UIWindow *window;
    ReturnMeToViewController *viewController;
}

@property (nonatomic, retain) IBOutlet UIWindow *window;
@property (nonatomic, retain) IBOutlet
        ReturnMeToViewController *viewController;

@end
```

Listing 6-2: **ReturnMeToAppDelegate.m**

```
#import "ReturnMeToAppDelegate.h"
#import "ReturnMeToViewController.h"

@implementation ReturnMeToAppDelegate

@synthesize window;
@synthesize viewController;

#pragma mark -
#pragma mark Application lifecycle

- (BOOL)application:(UIApplication *)application
        didFinishLaunchingWithOptions:
                          (NSDictionary *)launchOptions {
    .
    [window addSubview:viewController.view];
    [window makeKeyAndVisible];

    return YES;
}

- (void)applicationWillResignActive:
                          (UIApplication *)application {
}

- (void)applicationDidEnterBackground:
                          (UIApplication *)application {

    /*
}

- (void)applicationWillEnterForeground:
                          (UIApplication *)application {
}

- (void)applicationDidBecomeActive:
                          (UIApplication *)application {
}

- (void)applicationWillTerminate:
                          UIApplication *)application {
}

#pragma mark -
#pragma mark Memory management

- (void)applicationDidReceiveMemoryWarning:(UIApplication
        *)application {
}
```

```
-  (void)dealloc {
    [viewController release];
    [window release];
    [super dealloc];
}

@end
```

What this does is quite a bit; what it *doesn't* do is connect objects added to the user interface with the objects that need to know about them. How does my application access the phone number that the user enters, for example? Of course, I could root around in the code, but it would be much easier to have Interface Builder do it at application launch. I show you how that is (easily) done in the next chapter.

Your goal during startup should be to present your application's user interface as quickly as possible — quick initialization = happy users. Don't load large data structures that your application won't use right away. If your application requires time to load data from the network (or perform other tasks that take noticeable time), get your interface up and running first — and then launch the slow task on a background thread. Then you can display a progress indicator or other feedback to the user to indicate that your application is loading the necessary data or doing something important.

The application delegate object is usually derived from NSObject, the root class (the very base class from which all iPhone application objects are derived), although it can be an instance of any class you like, as long as it adopts the UIApplicationDelegate protocol. The methods of this protocol correspond to behaviors that are needed during the application lifecycle and are your way of implementing this custom behavior. Although you aren't required to implement all the methods of the UIApplicationDelegate protocol, every application should implement the following critical application tasks:

✔ Initialization, which I have just covered

✔ Responding to interruptions

✔ Responding to termination

✔ Responding to low memory warnings

I show you what has to be done to carry out these tasks in the last three sections.

Event processing

After a user launches your application, the functionality provided in the UIKit framework manages most of the application's infrastructure. Part of the initialization process mentioned in the preceding section involves setting up the main run loop and event-handling code, which is the responsibility of the UIApplication object.

When the application is onscreen, it's driven by external events — say, stubby fingers touching sleek screens, as shown in Figure 6-7. Here's a rundown of how external events drive an application:

1. **You have an event — the user taps a button, for example.**

   The touch of a finger (or lifting it from the screen) adds a touch event to the application's event queue, where it's *encapsulated* in — placed into — a UIEvent object. There's a UITouch object for each finger touching the screen so you can track individual touches. As the user manipulates the screen with his or her fingers, the system reports the changes for each finger in the corresponding UITouch object.

   My advice to you: Don't let your eyes glaze over here. This UIEvent and UITouch stuff is important, as you discover when I show you how to handle touch events while walking you through building the more advanced parts of the ReturnMeTo application.

2. **The run loop monitor dispatches the event.**

   When there is something to process, the event-handling code of the UIApplication processes touch events by dispatching them to the appropriate *responder* object — the object that has signed up to take responsibility for doing something when an event happens (when the user touches the screen, for example). Responder objects can include instances of UIApplication, UIWindow, UIView, and its subclasses (all which inherit from UIResponder).

3. **A responder object decides how to handle the event.**

   For example, a touch event occurring in a button (view) will be delivered to the button object. The button handles the event by sending an action message to another object — in this case, the UIViewController object. Setting it up this way enables you to use standard button objects without having to muck about in their innards — just tell the button what method you want invoked in your view controller, and you're basically set.

   Processing the message may result in changes to a view, or a new view altogether, or some other kind of change in the user interface. When this happens, the view and graphics infrastructure takes over and processes the required drawing events.

4. You're sent back to the event loop.

After an event is handled or discarded, control passes back to the run loop. The run loop then processes the next event or puts the thread to sleep if there's nothing more for it to do.

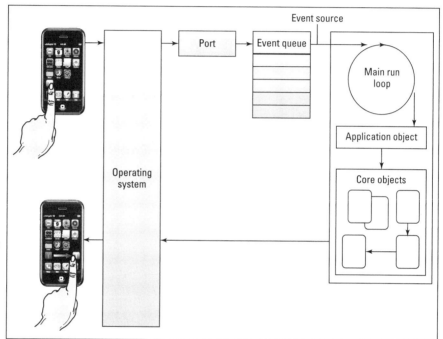

Figure 6-7: Processing events in the main run loop.

Normal processing is interrupted

Your application's normal processing can be interrupted in a few different ways:

- ✔ The application is interrupted by an incoming phone call, SMS, or calendar notification.
- ✔ The user presses the Sleep/Wake button.
- ✔ The user presses the Home button or the system launches another application, and the application is moved to the background.

In all these cases, your application first moves into the inactive state.

Your application becomes inactive

In all three cases, the sequence of events starts in the same way — with the `applicationWillResignActive:` message being sent to your application delegate. Using this method, you should pause ongoing tasks, disable timers, and generally put things on hold.

What happens after this depends on the nature of the interruption and/or how the user responds to the interruption. Your application may be

- ✔ Reactivated
- ✔ Moved to the background

The next two sections take a look at each scenario.

Your application is reactivated

If the user ignores the phone call, SMS, or calendar notification, the system sends your application delegate the `applicationDidBecomeActive:` message and resumes the delivery of touch events to your application.

If the user pressed the Sleep/Wake button, the system then puts the device to sleep. When the user wakes the device later, the system sends your application delegate the `applicationDidBecomeActive:` message, and your application receives events again. While the device is asleep, foreground and background applications do continue to run, but should do as little work as possible in order to preserve battery life.

In both cases, you can use the `applicationDidBecomeActive:` method to restore the application to the state it was in before the interruption. What you do depends on your application. In some applications, it makes sense to resume normal processing. In others — if you've paused a game, for example — you could leave the game paused until the user decides to resume play.

Your application moves to the background

When the user accepts the phone call, SMS, or calendar notification or presses the Home button or the system launches another application, your application then moves into the background, as shown in Figure 6-8.

In this case, your application will be sent the `applicationDidEnter Background:` message. In this method, you should include saving any unsaved data or key application *state* (what the user is looking at or doing in the application, such as the current view) to disk. (Okay, I know, the disk in the iPhone isn't really a disk; it's a solid state drive that Apple calls a disk, but if Apple calls it that, I probably should, too, just so I don't confuse too many people.) You need to do everything necessary to restore your application in case it's subsequently purged from memory as well as additional cleanup operations, such as deleting temporary files. Even though your application enters the background state, there is no guarantee that it will remain there indefinitely. If memory becomes a problem, for example, the OS will purge your background applications to make more room.

If your application is suspended when your application is purged, *it receives no notice that it is removed from memory.* You need to save any data beforehand!

If your application requests more execution time or it has declared that it does background execution, it is allowed to continue running after the `applicationDidEnterBackground:` method returns. If not, your (now) background application is moved to the *suspended* state shortly after returning from the `applicationDidEnterBackground:` method.

In the ReturnMeTo application, the only thing you're concerned with is saving the user's phone number. In the iPhoneTravel411 application you develop later in this book, you do background processing, and I explain what needs to be done in these methods to allow that to happen.

When your delegate is sent the `applicationDidEnterBackground:` method, your application has approximately five seconds to finish things up. If the method doesn't return before time runs out (or request more execution time from the OS), your application is terminated and purged from memory.

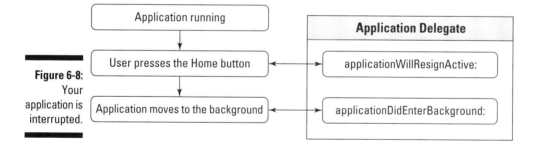

Figure 6-8: Your application is interrupted.

Your application resumes processing

At some point, it's likely that the user will once again want to use your application — the one that has been patiently sitting in background waiting for this opportunity. When this happens, your application delegate is sent the `applicationWillEnterForeground:` and `applicationDidBecome Active:` messages. This is shown in Figure 6-9.

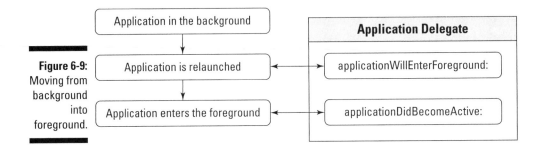

Figure 6-9:
Moving from
background
into
foreground.

I explain in the previous section what you need to do in the `application DidBecomeActive:` method — restart anything it stopped doing and get ready to handle events again.

While an application is suspended, the world still moves on, and the OS tracks all the things the user is doing that may impact your application. For example, the user may change the device orientation from landscape to portrait to landscape and then portrait. Although the OS will keep track of all events, it will only send you a single event that reflects the final change you need to process — in this case, the change to portrait. I won't cover this particular scenario in the ReturnMeTo application, but I discuss it briefly when I show you how to do background processing on the iPhoneTravel411 application in Chapter 18.

The way this is handled is very different than in OS 3.x. Now, just to show you how up-to-date a guy I am, this book is showing you how to create its applications by using OS 4.0. But what if you want your application to be able to run under both OS 3.x and OS 4.0? Doing that is beyond the scope of this book, but if you go to my Web site (`www.nealgoldstein.com`), you can find a short tutorial on how to change the ReturnMeTo application to run under both OS 3.x and OS 4.0.

Application termination

Whereas in OS 4.0 applications are generally moved to the background when

✔ The application is interrupted by an incoming phone call, SMS, or calendar notification.

✔ The user presses the Sleep/Wake button.

✔ The user presses the Home button or the system launches another application, and the application is moved to background.

. . . that isn't always the case. Under the following circumstances, your application is terminated if any of those things happen:

- ✔ The application is compiled by using the iPhone SDK 3.x or earlier.
- ✔ The application is running on a device running iPhone OS 3.x or earlier.
- ✔ The current device doesn't support multitasking.
- ✔ You decide you don't want your application to run in the background and you set the `UIApplicationExitsOnSuspend` key in its `Info.plist` file.

Your application delegate is sent the `applicationWillTerminate:` message and you do what you need to do to terminate your application, including saving any unsaved data or key application state, as well as additional cleanup operations, such as deleting temporary files. This is the same thing you would have done in the application method `applicationDidEnter Background:`.

It's worth noting (as of July 2010) that in the last two cases, before the application is terminated, and the `applicationWillTerminate:` message sent, the `applicationDidEnterBackground:` message is also sent. This means that for applications you compile and run only under iOS 4 and beyond, you only have to do all that cleanup and file saving in `applicationDid EnterBackground:`.

Your `applicationWillTerminate:` method implementation has about five seconds to do what it needs to do and return. Any longer than that and your application is terminated and purged from memory. (Those guys don't kid around.)

Even if you develop your application by using the iPhone OS 4.0 SDK and later — as you will be doing, so you stay as up-to-date as me — you must still be prepared for your application to be terminated. If memory becomes an issue (and it inevitably will if there are enough applications in the background), the system might remove your application from memory in order to make more room. If it does remove your *suspended* application, *it does not give you any warning, much less notice!* However, if your application is currently running in the background (I explain how that can happen in Chapter 18), the system does call the `applicationWillTerminate:` method of the application delegate.

The Managed Memory Model Design Pattern

You may remember that I mention in Chapter 2 that there's one other design pattern I explain here in Chapter 6 — how your application has to come to terms with memory management.

One of the main responsibilities of all good little applications is to deal with low memory. So, the first line of defense is (obviously) to understand how you as a programmer can help them *avoid* getting into that state.

In the iPhone OS, each program uses the virtual-memory mechanism found in all modern operating systems. But virtual memory is limited to the amount of physical memory available. This is because the iPhone OS doesn't store changeable memory (such as object data) on the disk to free up space and then read it in later when it's needed. Instead, the iPhone OS tries to give the running application the memory it needs — using memory pages that contain read-only content (such as code), where all it has to do is load the original read-only content back into memory when it's needed. Of course, this may be only a temporary fix if those resources are needed again a short time later.

If memory continues to be limited, the system may also send notifications to the running application, asking it to free up additional memory. This is one of the critical events that all applications must respond to.

Observing low-memory warnings

When the system dispatches a low-memory notification to your application, it's something you must pay attention to. If you don't, you're inviting disaster. (Think of your low-fuel light going on as you approach a sign that says "Next services 100 miles.") `UIKit` provides several ways of setting up your application so that you receive timely low-memory notifications:

- ✔ Implement the `applicationDidReceiveMemoryWarning:` method of your application delegate. Your application delegate can then release any data structure or objects it owns — or notify the objects to release memory they own.

- ✔ Override the `didReceiveMemoryWarning:` method in your custom `UIViewController` subclass. The view controller could then release views — or even other view controllers — that are off-screen.

- ✔ Register to receive the `UIApplicationDidReceiveMemoryWarning Notification:` notification. A model object can then release data structures or objects it owns that it doesn't need immediately and can re-create later.

Each of these strategies gives a different part of your application a chance to free up the memory it no longer needs (or doesn't need right now). As for how you actually get these strategies working for you, this is something that is dependent on your application's architecture. That means you need to explore it on your own.

Not freeing up enough memory will result in the iPhone's OS sending your *active* application the `applicationWillTerminate:` message and shutting you down. For many applications, though, the best defense is a good offense, and you need to manage your memory effectively and eliminate any memory leaks in your code.

Avoiding the warnings

When you create an object — a window or button for example — memory is allocated to hold that object's data. The more objects you create, the more memory you use, and the less there is available for additional objects you might need. Obviously, it's important to make available (or *de-allocate*) the memory that an object was using when the object is no longer needed. This is what people mean by *memory management.*

The Objective-C language — the application-programming language used to develop iPhone applications — uses reference counting to figure out when to release the memory allocated to an object. It's your responsibility (as a programmer) to keep the memory-management system informed when an object is no longer needed.

Reference counting is a pretty simple concept. When you create the object, it's given a reference count of 1. As other objects use this object, they use methods to increase the reference count and decrease it when they're done. When the reference count reaches 0, the object is no longer needed, and the memory is de-allocated.

Some basic memory-management rules you shouldn't forget

Here are the fundamental rules when it comes to memory management:

- ✔ Any object you create by using `alloc` or `new`, any object that contains `copy`, and any object you send a `retain` message to is yours — you own it. That means you're responsible for telling the memory-management system when you no longer need the object and that its memory can now be used elsewhere.

- ✔ Within a given block of code, the number of times you use `new`, `copy`, `alloc`, and `retain` should equal the number of times you use `release` and `autorelease`. You should think of memory management as consisting of pairs of messages. If you balance every `alloc` and every `retain` with a `release`, your object will eventually be freed up when you're done with it.

✔ When you assign an instance variable by using an accessor with a property attribute of `retain`, `retain` is automatically invoked — that is, you now own the object. Implement a `dealloc` method to release the instance variables you own.

✔ Objects created any other way (through convenience constructors or other accessor methods) are not your problem.

If you have a solid background in Objective-C memory management (all three of you out there), these instructions should be straightforward or even obvious. If you don't have that background, no sweat: I show you how to do this in practice and point out the pitfalls as I present the code for the ReturnMeTo application.

Reread this section!

Okay, there are some aspects of programming that you can skate right past, without understanding what's really going on, and still create a decent iPhone application. But memory management is *not* one of them!

There is a direct correlation between the amount of free memory available and your application's performance. If the memory available to your application dwindles far enough, the system will be forced to terminate your application. To avoid such a fate, keep a few words of wisdom in mind:

✔ Make minimizing the amount of memory you use a built-in feature of your implementation design.

✔ Be sure to use the memory-management functions I explain as I develop the ReturnMeTo application.

✔ In other words, be sure to clean up after yourself, or the system will do it for you, and it won't be a pretty picture.

Whew!

Congratulations — you have just gone through the *Classic Comics* version of another several hundred pages of Apple documentation, reference manuals, and how-to guides.

Although there's a lot left unsaid (though less than you might suspect), what's in this chapter is enough to not only to get you started but also to keep you going as you develop your own iPhone applications. It provides a frame of reference on which you can hang the concepts I throw around with abandon in upcoming chapters — as well as the groundwork for a deep enough understanding of the application lifecycle to give you a handle on the detailed documentation.

Now it's time to move on to the really fun stuff.

Part III
From "Gee, That's a Good Idea," to the App Store

The 5th Wave   By Rich Tennant

"We're here to clean the code."

In this part . . .

You're not the only one who dreams of the riches and glory an iPhone application can bring — your author has dreams, too. In this part, I take you through the entire process of developing the small yet real and sort of functional application that you built the user interface for in Part II. Not far down the development trail, you get to see how to add all the features your app needs and even what to do when you change your mind (whether because you found a better way to do something or your users demanded it). I end up showing you how to get your application running on the iPhone and then into the App Store. All this should keep you out of trouble for a while, but at the end, you can start thinking about a trip around the world by private jet, or something equally reasonable.

Chapter 7

Actually Writing Code

*I*f you jumped right to this chapter, you're probably really itching to start writing some code. I understand the urge, and if you're up for it, this chapter — and the chapters that follow — can definitely scratch that code-writing itch for you. (If you start here and then find some of this tough going, you may want to jump back a chapter or two so I can fill you in on some concepts you need under your belt in order to make your coding experience a bit more productive — and more fun!)

Buckle Up, It's Time to Code

Previous chapters talk about design principles in general, as well as about the specific iPhone developer tools (Xcode, Interface Builder) available to you. Chapter 5, for example, has you create the skeleton for a fully functioning iPhone application (my ReturnMeTo jewel) — and now you get to flesh it all out with the code necessary for transforming the ReturnMeTo application from an app that just sits there and looks pretty to an app that actually *does* something.

A quick refresher peek at Chapter 5 shows you that quite a bit of the ReturnMeTo application is already in place and ready to go. If you click the text field, for example, you *do* get the keyboard — though you can't do anything with it just yet. Both the text field and the label are automatically created from the nib file — which is great — but somehow you have to accomplish two tasks:

✔ Get what the user enters into the text field.

✔ Display the user input in the label you created.

In effect, you have to connect things up so the left hand knows what the right hand is doing.

Fortunately, the framework was designed to allow you to do this easily and gracefully. The view controller can refer to objects created from the nib file by using a special kind of *instance variable* (a variable defined as part of a class, with each object of that class having its own copy) referred to as an *outlet.* If I want (for example) to be able to access the text field object in my ReturnMeTo application, I take two steps:

1. Declare an outlet in my code.

2. Use Interface Builder to point the outlet to the text field I created earlier.

Then when my application is initialized, the text field outlet is automatically initialized with a pointer to the text field. I can then use that outlet from within my code to get the text the user entered in the text field.

The fact that a connection between an object and its outlets exists is actually stored in a nib file. When the nib file is loaded, each connection is reconstituted and reestablished — thus enabling you to send messages to the object. IBOutlet is the keyword that tags an instance-variable declaration so the Interface Builder application knows that a particular instance variable *is* an outlet — and can then enable the connection to it with Xcode.

In my code, it turns out I need to create two outlets — one to point to the text field and one to point to the label where I display the number the user enters. To get this outlet business started, I need to *declare* it, which I do with the help of the aforementioned IBOutlet keyword.

Okay, I'm guessing you realize that *declaring* something in programming doesn't involve standing on a soapbox in Hyde Park and saying something at the top of your lungs. Declaring something code-wise involves . . . writing code. (You knew that.) More specifically, in iPhone application development, declaring something code-wise involves writing code by using the Xcode editor — which leads you right to the next section.

The Xcode Code Editor

The main tool you use to write code for an iPhone application is the Xcode text editor. Apple has gone out of its way to make the text editor as user friendly as possible, as evidenced by the following list of (quite convenient) features:

✔ **Code Sense:** As you type code, you can have the Editor help by inserting text that completes the name of whatever Xcode thinks you're going to enter.

Using Code Sense can be really useful, especially if you're like me and forget exactly what the arguments are for a function. When Code Sense is active (it is by default), Xcode uses the text you typed, as well as the context within which you typed it, to provide suggestions for completing what it thinks you're *going to* type. You can accept suggestions by pressing Tab or Return. You may also display a list of completions by pressing Escape.

✔ **Code Folding:** With Code Folding, you can collapse code that you're not working on and display only the code that requires your attention. You do this by clicking in the column to the left of the code you want to hide.

✔ **Switching between header and implementation windows:** On the toolbar above the code editor, click the last icon before the lock to switch from .h to .m (header and implementation), and vice versa. While the header lets you see the class's instance variables and method declarations, you find your actual code in the implementation file. If you look in the Groups & Files list of the Project window, you can see the separate .h and .m files for the four classes we have started with.

✔ **Launching a file in a separate window:** Double-click the filename to launch the file in a new window. If you have big monitors or multiple monitors, this new window lets you look at more than one file at a time. You could, for example, look at the method of one class *and* the method it invokes in the same, or even a different class.

Accessing Documentation

Like many developers, you may find yourself wanting to dig deeper when it comes to a particular bit of code. That's when you really appreciate Xcode's Quick Help, header file access, Documentation window, Help menu, and Find tools. With these tools, you can quickly access the documentation for a particular class, method, or property.

To see how this works, let's say I have the Project window open with the code displayed in Figure 7-1. What if I wanted to find out more about `UIApplicationDelegate`?

Quick Help

Quick Help is an unobtrusive window that provides the documentation for a single symbol. It pops up inline, although you can use Quick Help as a symbol inspector (which stays open) by moving the window after it opens. You can also customize the display in Documentation preferences in Xcode preferences.

You can get Quick Help for a symbol by pressing Option and then double-clicking the symbol in the Text Editor (in this case, UIApplicationDelegate; see Figure 7-1).

Figure 7-1:
Getting
Quick Help.

The header file for a symbol

Headers are a big deal in code because they're the place where you find the class declaration, which includes all of its instance variables and method declarations. To get the header file for a symbol, press ⌘ and then double-click the symbol in the Text Editor. For example, see Figure 7-2, where I pressed ⌘ and then double-clicked UIApplicationDelegate.

This works for the classes you create as well.

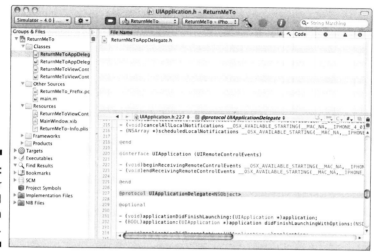

Figure 7-2:
The header
file for UI
Application
Delegate.

Documentation window

The Documentation window lets you browse and search items that are part of the iPhone Reference Library as well as any third-party documentation you have installed.

You access the documentation by pressing ⌘+Option+double-clicking a symbol to get access to an Application Programming Interface (better known as API) reference (among other things) that provides information about the symbol. This enables you to get the documentation about a method to find out more about it or the methods and properties in a framework class. In Figure 7-3, I pressed ⌘+Option+double-clicked UIApplicationDelegate.

Using the Documentation window, you can browse and search the developer documentation — the API references, guides, and article collections about particular tools or technologies — installed on your computer.

It's the go-to place for getting documentation about a method or more info about the methods and properties in a framework class. (Using the API reference is how I discovered that a view has a frame, which I use in Chapter 8 to determine how much to scroll the view to keep the text field visible.)

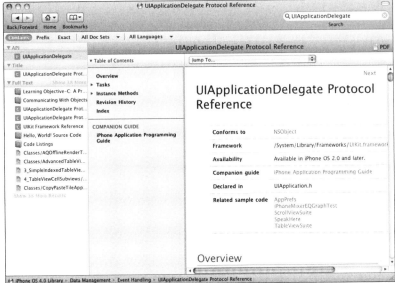

UIApplicationDelegate Protocol Reference

Figure 7-3:
The Docu-
mentation
window.

Help menu

The Help menu search field lets you search Xcode documentation as well as open the Documentation window and Quick Help.

You can also right-click a symbol and get a pop-up menu that gives you similar options to what you see in the Help menu (and other related functions). This pop-up menu is shown in Figure 7-4.

Find

Xcode can also help you find things in your own project. The submenu that you can access by choosing Edit⇨Find provides several options for finding text in your own project.

As your classes get bigger, sometimes you'll want to find a single symbol or all occurrences of a symbol in a file or class. You can easily do that by choosing Edit⇨Find⇨Find or by pressing ⌘+F, which opens a Find toolbar to help you search the file in the Editor window. In Figure 7-5, for example, I typed **viewDidLoad** in the Find toolbar, and Xcode found all the instances of viewDidLoad in that file and highlighted them for me.

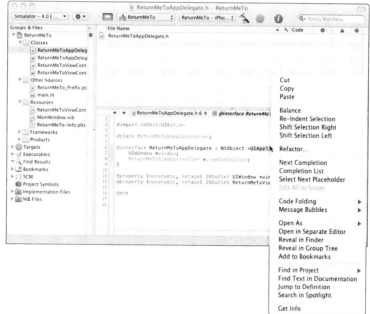

Figure 7-4:
Right-click
UI
Application
Delegate.

Figure 7-5:
Finding view
DidLoad in
a file.

You can also use Find to go through your whole project by choosing Edit➪
Find➪Find in Project or by pressing ⌘+Shift+F. I pressed ⌘+Shift+F, which
opened the window shown in Figure 7-6. I typed **ReturnMeToViewController**,
and then I selected In Project from the drop-down menu. You can specify in

what sets of files (open project files, and so on) you want to search. (A great feature for tracking down something in your code — you're sure to use it often.)

If you select a line in the top pane, as you can see in Figure 7-6, the file in which that instance occurs is opened in the bottom pane and the reference highlighted.

You can also narrow down your search in both the Find and Project Find windows. In the Find toolbar in Figure 7-7, select the little triangle to the right of the magnifying glass; in the Project Find window (refer to Figure 7-6) you can use the drop-down menus and check box in the window's top pane. You have quite a few options, including searches using whole words, case, and the ability to use regular expressions (yet another language like Objective-C, but one that is interpreted by Xcode and allows you to do very sophisticated searches). I leave it to you to explore this feature on your own.

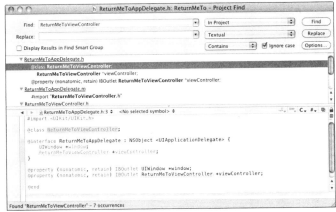

Figure 7-6:
Project Find.

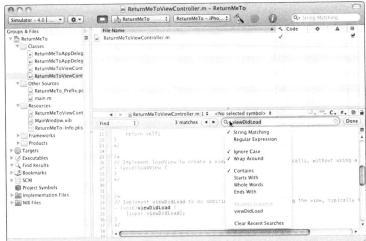

Figure 7-7:
Narrowing down your search.

Adding Outlets to the View Controller

Now that you have some idea of how to use the Xcode editor, it's time to write some code. Before taking you on the editor tour, I mention that one of the things I needed to do was add outlets to my ReturnMeTo application. That's what you're going to do now — add outlets to the ReturnMeToViewController. Here's how:

1. **Go to the Xcode Project window and, in the Groups & Files list, click the triangle next to Classes to expand the folder.**

2. **From the Classes folder, select ReturnMeToViewController.h — the header file for ReturnMeToViewController.**

   The contents of the file appear in the main display pane of the Xcode editor, as shown in Figure 7-8. (Of course yours won't have all that code in it yet — you enter it in Step 4.)

Figure 7-8:
ReturnMe
ToView
Controller.h.

3. **Look for the following lines of code in the header:**

   ```
   #import <UIKit/UIKit.h>

   @interface ReturnMeToViewController: UIViewController{

   }
   @end
   ```

   Got it? Great.

4. **Type the following lines of code between UIViewController{ and @end. (The curly brace you see below the last IBOutlet and first @property statements is already there.)**

```
IBOutlet UITextField *textField;
IBOutlet UILabel *label;
}
@property (nonatomic, retain) UITextField *textField;
@property (nonatomic, retain) UILabel *label;
```

When you're done typing, your code should look exactly like Figure 7-8.

The first two lines of code you added here declare the outlets, which will automatically be initialized with a pointer to the text field (`textField`) and label objects (`label`) when the application is launched. But while this will happen automatically, it won't *automatically* happen automatically. I have to help it out a bit.

In procedural programming, variables are generally fair game for all. But in object-oriented programming, a class's instance variables are tucked away inside an object and shouldn't be accessed directly. The only way for them to be initialized is for you to create what are called *accessor methods,* which allow the specific instance variable of an object to be read and (if you want) updated. Creating accessor methods is a two-step process that begins with a `@property` declaration, which tells the compiler that there *are* accessor methods.

And that is what I do with the code in this step; I code corresponding `@property` declarations for each `IBOutlet` declaration. Notice the arguments to the `@property` declaration. These specify how the accessor methods are to behave — I explain exactly what that means in the next section. For now, just know that you need to add them.

Although I'm using properties here, they are really not necessary. I could access the outlets directly. I do it because it's the way it's done in quite a bit of the Apple sample code.

5. **Go back to the Classes folder in the Groups & Files list and select** `ReturnMeToViewController.m` — **the implementation file for** `ReturnMeToViewController`.

6. **Look for the following lines of code in the implementation file:**

```
#import "ReturnMeToViewController.h"

@implementation ReturnMeToViewController
```

They're pretty much right at the top.

7. **Type the following two lines of code after** `@implementation` `ReturnMeToViewController` **and before anything else.**

```
@synthesize textField;
@synthesize label;
```

When you're done, your code should like what you see in Figure 7-9.

Although the `@property` declaration tells the compiler that there are accessor methods, they still have to be created. In the good-old days, you had to code these accessor methods yourself and, in a large program, it got to be very tedious. Fortunately, Objective-C creates these accessor methods for you whenever you include an `@synthesize` statement for a property.

That's what you do with the code in this step. The two `@synthesize` statements tell the compiler to create two accessor methods for you — one for each `@property` declaration.

8. **Scroll down the code for `ReturnMeToViewController.m` until you reach the following lines:**

```
- (void)dealloc {

    [super dealloc];
}
```

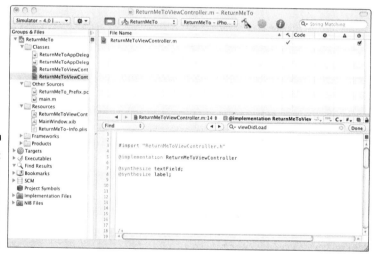

Figure 7-9:
Completing the addition of the accessors.

You can use ⌘+F to find something in a single file, as opposed to Shift+⌘+F, which finds it in all project files.

9. **Enter the following two lines of code between the - (void)dealloc { and [super dealloc]; lines:**

```
[textField release];
[label release];
```

The new code should look like what you see in Figure 7-10.

If you remember my obsession with memory management from previous chapters, you can likely recognize `release` as a tool for freeing up no-longer-needed memory commitments. (And if you haven't yet heard my memory-management stump speech, fear not: You get a chance to hear it later in this chapter.)

That's it. You've added outlets to your view controller. Step back and admire your handiwork. Then move on to the next section and see how the little snippets of code you added above to your `ReturnMeToViewController.m` and `ReturnMeToViewController.h` files tie in with the basic principles of programming, using the Objective-C language.

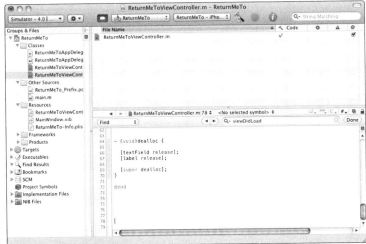

Figure 7-10:
Doing a little
memory
manage-
ment.

Objective-C properties

As you soon discover, you're going to use properties *a lot.* In the preceding section, I had you blindly follow me and add the properties. But by now you probably figured out that I don't believe you should be doing things blindly, so in this section, I get to explain what you need to know about properties.

Now, you may remember that, in object-oriented programming, a class's instance variables are tucked away inside an object and shouldn't be accessed directly. If you need to have an instance variable accessible by other objects in your program, you need to create accessor methods for that particular instance variable. (This will sound familiar from the previous section.)

For example, in Chapter 9, you're going to add an instance variable `saved Number`, to the `ReturnMeToAppDelegate`. You do that because you need something to hold the telephone number someone's supposed to use to call you when he or she finds your lost iPhone. The `ReturnMeToAppDelegate` saves that number when the application terminates, and loads it when it launches. But the `ReturnMeToViewController` needs access to that number to display it in the view, and needs to update it when the user enters a new number.

The methods that provide access to the instance variables of an object are called *accessor methods,* and they effectively get (using a *getter method*) and set (using a *setter method*) the values for an instance variable. Although you could code those methods yourself, it can be rather tedious. This is where properties come in. The Objective-C Declared Properties feature provides a simple way to declare and implement an object's accessor methods. The compiler can synthesize accessor methods for you, according to the way you told it to in the property declaration.

Objective-C creates the getter and setter methods for you by using an `@property` declaration in the interface file, combined with the `@synthesize` declaration in the implementation file. The default names for the getter and setter methods associated with a property are *whateverTheProperty NameIs*: for the getter and set*WhateverThePropertyNameIs*: for the setter. (You replace what's in italic with the actual property name.) For example, the accessors generated in the ReturnMeTo application are `text Field` as the getter and `setTextField`: as the setter. Similarly, the names for the label accessors are `label` and `setLabel`: for the getter and setter, respectively.

All that being said, at the end of the day, you need to do three things in your code to have the compiler create accessors for you:

1. **Declare an instance variable in the interface file.**

2. **Add an `@property` declaration of that instance variable in the same interface file (usually with attributes `nonatomic` and `retain`).**

   This is what you do in Step 4 in the preceding section. The declaration specifies the name and type of the property and some attributes that provide the compiler with information about how exactly you want the accessor methods to be implemented.

   For example, the declaration

   ```
   @property (nonatomic, retain) UITextField *textField;
   ```

   declares a property named `textField`, which is a pointer to a `UITextField` object. As for the two attributes, `nonatomic` and `retain`, `nonatomic` tells the compiler to create an accessor to return the value directly, which is another way of saying that the accessors can be interrupted while in use. (This works fine for applications like this one.)

   The second value, `retain`, tells the compiler to create an access method that sends a `retain` message to any object that is assigned to this property. This will keep it from being *de-allocated* — having its memory taken back by the iPhone OS to use elsewhere — while you're still using it. (I go into de-allocation a bit more when I explain the `dealloc` method in the "Memory management" section, coming up next.)

3. **Use @synthesize in the implementation file so that Objective-C generates the accessors for you.**

   The @property declaration (like the two you placed in the interface file in Step 4 in the preceding section) only declares that there are accessors. It's the @synthesize statement (like the two you place in the implementation file in Step 7 in the preceding section) that tells the compiler to create them for you. Using @synthesize results in four new methods.

   ```
   textField
   setTextField:
   label
   setLabel:
   ```

If I didn't use @synthesize, it would be up to me to implement the methods myself, using the attributes in the @property statement. So if I were to write my own accessors, I would be responsible for sending a retain message to the textField or label when it is assigned to the instance variables. While there are circumstances when you do want to do that, I don't get into them in this book.

Memory management

In Chapter 6, I spend a lot of time nagging you about memory management — and I promised there that I would make sure to show you how to do this memory-management thing, using a real-world example. I start to keep that promise by explaining the dealloc method created for me by the Xcode template when I created the ReturnMeTo application. Now, recall that in Step 9 in the earlier section, "Adding Outlets to the View Controller," I asked you to add

```
[textField release];
[label release];
```

to the dealloc method, so the end product looks like the following:

```
- (void)dealloc {

   [textField release];
   [label release];

   [super dealloc];
}
```

Well, here's why: Adding these bits of code *releases* textField and label.

Chapter 6 gives you some handy memory-management rules. Here's one of them:

You own an object you create with `alloc` or `new` or if it contains `copy` or if you send it a `retain` message. That means you're responsible for telling the memory-management system you're done with it.

In other words, you have to release it when you're done.

So why, then, you might ask, do I have to release `textField` and `label`? If I had created the `textField` and `label` by using `alloc` or `new`, obviously it would be my job to release them. But I didn't do that — or did I?

No, I didn't, but what I *did* do was send a `retain` message to both `textField` and `label`.

"Oh, yeah?" "Where?" you might ask.

Check out the `@property` declarations you made earlier:

```
@property (nonatomic, retain) UITextField *textField;
@property (nonatomic, retain) UILabel *label;
```

You see, lo and behold, `retain`.

The impact of adding that simple two-syllable word `retain` is that any time I assign to that instance variable — assigning a phone number to a label, for example — a `retain` message will be sent to it, and that is precisely what happens at runtime for the outlets.

Notice that while the `dealloc` method generated by Xcode for the `ReturnMeToViewController` only invoked `[super dealloc]`, the one generated for me in `ReturnMeToAppDelegate` did more than that.

```
- (void)dealloc {
  [viewController release];
  [window release];
  [super dealloc];
}
```

This is because even though Xcode didn't know about the `label` and `text Field` that I was going to create (it can't read my mind after all), it did know about the window and view controller it created for me, and it created the code to release them in the `dealloc` method.

Connecting the Pieces in Interface Builder

Earlier in this chapter, I tell you that if you want to be able to access the label and text field objects in my ReturnMeTo application, you have to take two steps:

1. Declare an outlet in your code.

2. Use Interface Builder to point the outlet to the label and text fields you created earlier in Interface Builder.

You created the outlets and their accessor methods in your code — I saw you do it in the previous section. Now I'm going to show you how to create the connection in Interface Builder so that when the nib file is loaded, the nib loading code will create these connections automatically, using the *accessors you had the compiler create for the* `label` *and* `textField`. (Aha!) With these connections established, you'll be able to send messages to your interface objects. (I show you how to *receive* messages from interface objects in Chapter 8.)

So, it's connection time.

1. **For your ReturnMeTo project, be sure to add the instance variables, `@property` declaration, and `@synthesize` statement to your code as spelled out in Steps 4 and 7 in the "Adding Outlets to the View Controller" section, earlier in this chapter.**

2. **Choose File⇨Save or press ⌘+S to save what you have done for each file.**

   You have to save your code; otherwise, Interface Builder won't be able to find it.

3. **In the Project window, double-click `ReturnMeToViewController. xib` in the Groups & Files list to launch Interface Builder. (You'll find the `.xib` file in the Resources folder.)**

   Interface Builder duly makes an appearance on-screen, with the main nib window and the View window open for inspection. (For more on the mechanics of Interface Builder, see Chapter 5.)

4. **Holding down the Control key, click the File's Owner icon in the main nib window and drag it to the label field in the View window, as shown in Figure 7-11.**

   You need to use the Control key here, or you just end up dragging the File's Owner icon, rather than initiating a connection.

You should see the `label` value (650 555 1212) appear when your cursor is over the label field. (If you remember, that was the value you set when you created the label in Chapter 5.)

5. **Choose `label` from the pop-up menu.**

   Interface Builder now knows that one of the File Owner's outlets (in this case, the `label` value you select in the pop-up menu, which is one of the `ReturnMeToViewController` outlets) should point to the label at runtime.

Figure 7-11:
Dragging from the File's Owner to the label.

6. **With the cursor still over the label field, let go of the mouse button.**

   A pop-up menu appears, looking like the one in Figure 7-12.

Figure 7-12:
The label option.

There's another way to do this, however; one that's a little more obvious. To see how that method works, check out how I connect the other outlet — the `textField` outlet — to its text field.

7. Right-click the File's Owner icon in the main nib window to call up a dialog displaying a list of connections.

This particular dialog can also be accessed by choosing the Connections tab in the Interface Builder Inspector.

8. Drag from the `textField` outlet item in the dialog onto the text field in the View window, as shown in Figure 7-13.

Interface Builder now knows that the `textField` outlet should point to the text field at runtime. All is right with the world.

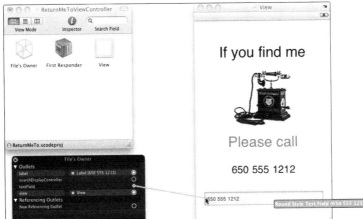

Figure 7-13: Connecting the text field in another way.

9. Go back to the Xcode Project window and click the Build and Run button to compile and build the application.

Figure 7-14 shows what happens if you click in the text field. Neat, huh?

The only problem, of course, is that after the keyboard comes up, you can't dismiss it and therefore can't see anything behind it. A bit of a problem, I admit, but it's easily fixable — as the next chapter makes clear.

Figure 7-14:
So you get a
keyboard.

Chapter 8

Entering and Managing Data

. . . .

. . . .

*T*hings aren't perfect with the ReturnMeTo application yet. The iPhone on the left in Figure 8-1 shows what happens when you select the text field to enter a phone number.

Not very useful, is it? Reminds me of the time I sat behind Yao Ming in the movie theater. (Can you say, "Down in front!"?)

The iPhone on the right in Figure 8-1 does it the way it should be — as in, the view moves down so you can see what you're typing.

Okay, maybe the movie-theater metaphor isn't that appropriate. (I was kidding about famous basketball players anyway.) It's not that the keyboard ducks down so you can better see the content view; rather, the idea is to *scroll the content view up* when the keyboard appears so you can see the phone number you're entering as you continue to develop the application.

Incidentally, implementing this little scrolling business is probably the most complicated thing you'll be doing with this application. It involves a number of different objects and methods that are invoked over the life of the application.

Figure 8-1:
Well, you do
get a
keyboard,
but not
much else.

I have an ulterior motive here: As I take you through implementing the scrolling of the content view, I also show you the dynamic flow of the application. Keeping a close eye on this flow will give you a good working sense of how and where to insert your own code to handle such tasks as these:

- Initialization and termination of your application
- Initialization and termination of views
- Processing user touches on the screen

Developing this knack is vital for your own applications. Fortunately, even the simple example in this chapter provides a structure that you're sure to use in much more sophisticated applications.

Scrolling the View

On iPhone applications, when a user touches a text field used for data entry, the keyboard scrolls up from the bottom of the screen. That's all fine and dandy. The problem is that, by default, if the text field you specified as the User Entry field is toward the bottom of the content view, that magically appearing keyboard is going to scroll up and cover every inch of the text field so the user can't see what he or she is entering. The solution (as I mention earlier; displayed on the right in Figure 8-1) is to scroll the content view — which includes the text field — up so the text field will still be visible.

Simple enough concept. Getting to it requires a number of steps:

1. Registering to be notified when the keyboard appears.

   This involves asking the iPhone OS to invoke a method I specify whenever the keyboard is about to scroll into view.

2. Deciding whether the text field will in fact end up being covered by the keyboard.

3. Moving the content view up so the keyboard will not cover the text field.

4. After the user is done editing, dismissing the keyboard and restoring the content view back to where it was.

5. Unregistering for keyboard notifications when the view is dismissed or not visible.

These steps are illustrated in Figure 8-2.

But before I get into any of that, you have to know where to put your code so it's invoked at the right time.

This is the scenario at runtime — although, as I explain shortly, this isn't the order in which I necessarily want to implement the code. I start with the code that does all the work, and then I show you registering and unregistering for keyboard notifications and how to get that code executed.

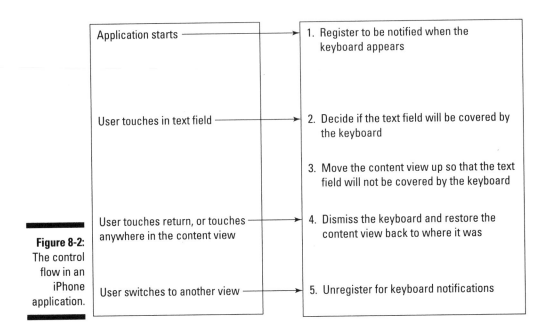

Application starts

Figure 8-2:
The control
flow in an
iPhone
application.

Where Does My Code Go?

One of the biggest challenges facing a developer working with a new framework is understanding where in the *control flow* — the sequence in which messages are sent during execution — he or she needs to add the code to get something done. I discuss some of these more traffic-cop-ish aspects of application development in Chapter 6 ("okay, your turn — now you over there, yield — now you in the right lane, go"), but I want to expand upon that discussion here, using the scrolling of a view in response to user action as a concrete example.

Figure 8-3 illustrates the higher-level control flow within our ReturnMeTo application. Look closely and you can see two objects — ReturnMeTo AppDelegate and ReturnMeToViewController — that you'd find at runtime if you went behind the screen and sifted through memory. (I bolded the two objects in the figure to make them stand out.) The next two sections look at what these two objects do for you as part of the ReturnMeTo application.

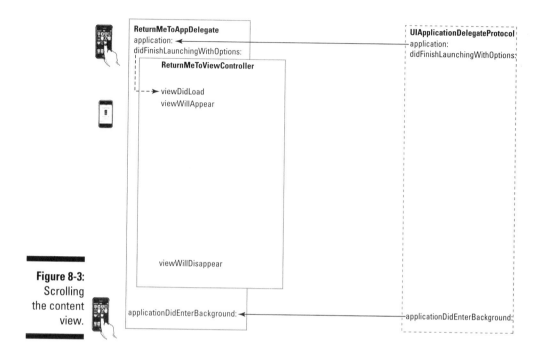

Figure 8-3:
Scrolling
the content
view.

The delegate object

The first, `ReturnMeToAppDelegate`, has two methods I want to call your attention to:

- ✔ `application:didFinishLaunchingWithOptions:`
- ✔ `applicationDidEnterBackground:`

The `application:didFinishLaunchingWithOptions:` message is sent at the very beginning of the application, before the user can even see anything on the screen. Here's where you insert your code to initialize your application — where you load data, for example, or restore the state of the application to where it was the last time the user exited.

In the case of the ReturnMeTo application, there really is only one state you need to concern yourself with, so you don't have to worry about saving any state. (I wouldn't count the keyboard scrolled up in full view as a state I'd want to leave the application in.) In more complex applications, you would have to work a bit in the `application:didFinishLaunchingWithOptions:` method to set things up correctly. I *do* show you how to both save and restore the state in Chapter 15, but even for a simple application like this, saving and restoring data is important.

As I explain in Chapter 6, the `applicationDidEnterBackground:` message is sent right before your application moves to the background, and it's the place where you store any unsaved data and save the current state of the application. If you check out Chapter 9, you can see how I use `application DidEnterBackground:` to save the phone number the user entered.

This code only runs correctly with iPhone OS 4.0 or later. In case you do want your code to run under iPhone OS 3.x, I've posted a short tutorial on how to change the ReturnMeTo application to run under both OS 3.x and OS 4.0 on my Web site (`www.nealgoldstein.com`).

The `application:didFinishLaunchingWithOptions:` method and the `applicationDidEnterBackground:` method are in the `UIApplicationDelegate` protocol. As I explain in Chapter 2, protocols are simply rules that spell out methods that can be implemented by any class. My `ReturnMeToAppDelegate` (for example) has adopted the `UIApplicationDelegate` protocol, so if I implement `application DidFinishLaunching:` and `applicationDidEnterBackground:` in the `ReturnMeToAppDelegate` implementation, they will be automatically invoked.

The controller object

`ReturnMeToViewController` is the controller responsible for managing the application's *view* — what the user sees and interacts with, including text editing. There are three methods here I want to call your attention to:

- ✔ `viewDidLoad`
- ✔ `viewWillAppear:`
- ✔ `viewWillDisappear:`

The `viewDidLoad` message is sent right after the view has been loaded from the nib file. (Check out Chapter 6 for a complete explanation of that loading process.) This is the place where you insert your code for *view initialization,* which in this case means updating the text in the label to show the phone number.

The `viewWillAppear:` message is sent right before the view will appear. This is the place to insert your code to do anything needed before the view becomes visible.

Finally, the `viewWillDisappear:` message is sent right before a view is dismissed or covered up. This is the place to insert your code to do anything you need to do before a view may be released or freed.

These three methods are declared in the `UIViewController` class and are invoked at the appropriate times by the framework. In this case, because `ReturnMeToViewController` is derived from the `UIViewController` class, I need to *override* those methods. To do that, I simply implement a new method with the same name as one defined in the `UIViewController` class in the `ReturnMeToViewController` implementation.

Most of the time, you do your initialization at the application level by using the `application:didFinishLaunchingWithOptions:` method in your application delegate. As for your initialization of the view level, you normally take care of that by using the `viewDidLoad` and `viewWillAppear:` methods in your `UIViewController` derived class — `ReturnMeToViewController`, in this example. When it comes to shutting down your app, you use the `applicationDidEnterBackground:` method in your application delegate to handle all the chores of putting your application on hold while it hangs out in the background — and the `viewWillDisappear:` method in your `UIViewController` to take care of what needs to be done when the view is dismissed.

Knowing the control flow I outline in this section — how and where to insert your code in order to add your specific application functionality — makes developing your own application much easier. Trust me on that one.

Where Where Where

Time to get your hands dirty again. Listings 8-1 and 8-2 show the code pulled together in Chapter 7 as the foundation for our ReturnMeTo application. (The bolded code is the code you actually had to type in, whereas the unbolded code was already put in place for you by the `UIKit` framework.)

Listing 8-1:   ReturnMeToViewController.h

```
#import <UIKit/UIKit.h>

@interface ReturnMeToViewController: UIViewController {

  IBOutlet UITextField *textField;
  IBOutlet UILabel     *label;

}

@property (nonatomic, retain) UITextField *textField;
@property (nonatomic, retain) UILabel *label;

@end
```

Listing 8-2:   ReturnMeToViewController.m

```
#import "ReturnMeToViewController.h"

@implementation ReturnMeToViewController

@synthesize textField;
@synthesize label;


- (void)didReceiveMemoryWarning {

  [super didReceiveMemoryWarning];
}

- (void)dealloc {

  [textField release];
  [label release];
  [super dealloc];
}

@end
```

To enable scrolling of the content so that all content will be accessible at all times, you need to make a few changes to both the ReturnMeToView Controller.h file (the file containing interface stuff, as spelled out in Listing 8-1) and the ReturnMeToViewController.m file (the file containing implementation stuff as well as a few other things whose purpose will become abundantly clear over the course of this chapter). Listing 8-2 shows the ReturnMeToViewController.m file as it stands right now.

Building on a Foundation

The code put together in Chapter 7 for the ReturnMeTo application is a great start, but you still have to address the My Keyboard Is Hiding Crucial Parts of My View problem I mention at the beginning of this chapter. Now, the solution clearly is to move things around so that the iPhone keyboard no longer hogs the view, but getting to that point is a bit tricky. For example, if I'm going to scroll the content view when the keyboard appears, I need to know *when* the keyboard is going to appear. Luckily for me, the iPhone OS has something in place that fits my needs exactly. It's a *Notification* system.

Notification

Notification is a system that allows objects within an application to learn about changes that occur elsewhere in that application. Usually, objects get information by messages that come to them. But that means the object that sends the message must know what objects it needs to update whenever it does something that those objects care about. And face it, the UIWindow object, being as it is in the keyboard displaying business, has no clue about my ReturnMeToViewController object.

That's where notification comes in. Notification is a broadcast model where I can register my objects to be notified of a particular event. I can even post a notification, although I'm not interested in doing that here. Notifications are managed by a single object, NSNotificationCenter, which is accessed by using the class method defaultCenter.

In iOS 4.0, the notification system can be used in one of two ways to make sure the code that scrolls the view gets called.

The traditional way (and the only way in the pre-iOS 4 world) is to tell the notification center that I want it to send me a particular message when a UIKeyboardWillShowNotification is posted. That notification is posted by the UIWindow class. When this happens, I also receive a notification object (NSNotification), which has the information I'm going to be interested in.

The new modern way (in iOS 4 only) is to pass a Block object to the notification center and tell it to execute the block when the notification is posted. I explain Block objects in Chapter 2, and in this chapter I show you how to use them. Wanting to be all hip and modern, this is the approach I take.

For all of this to work correctly, however, timing is everything. So before I explain about the mechanics of scrolling the view, I want to explain where and how to register the notification.

Registering a notification

If you want to have your app do something in response to a user's action in a view — the user touching the text field and the keyboard appearing, to take the most obvious example — you want to make sure that you informed the notification center what's what before the user actually has a chance to even think about touching inside the text field.

At this point, take a look at Figure 8-4; it shows (among other things) that the best place for that to happen — that is, the best place to let the notification center know that you want your code executed — is *after* view initialization but *before* the view becomes visible.

I could (and usually would) do this in `viewDidLoad`, but `viewWillAppear:` is a handy place to know about because it's not only invoked the first time a view appears (as is the case with `viewDidLoad,`), but every other time as well. So, if in the `ReturnMeToViewController` you were to create a new view controller (`AnotherViewControllerToDoSomethingElse`), the `viewWillAppear:` message gets sent when the user dismisses `Another ViewControllerToDoSomethingElse` and returns to `ReturnMeToView Controller`. I explain this whole navigation mechanism again when you're working with the iPhoneTravel411 app in Chapter 14.

The iPhone icons on the left side of Figure 8-4 show that the view will appear after the `viewWillAppear:` method is invoked. That's the ideal place to insert the code that registers the fact that you want to be notified before the keyboard appears. All I need to do to override `viewWillAppear:` is add this method to my implementation, as spelled out in Listing 8-3. (For all the dirty details about adding your own code to that handy bunch of code provided by the frameworks, check out Chapter 7.)

Now, if you were to look in the `ReturnMeToViewController` code, guess what? You're not going to find `viewWillAppear:`, not as a stub, or even commented out. You're going to have to add it yourself. Instead of adding things at random, though, you may want to group similar code together, and I talk about how to do that later in this chapter. For right now, I'd put `view WillAppear:` right after the commented out `viewDidLoad`, which you'll use shortly. You can find that bit of code in the `ReturnMeToViewController.m` file by pressing ⌘+F, which opens up a Find toolbar that you can then use to find whatever you want in a particular file. Handy tool that one is.

If you look at Listing 8-3, you can see how I registered with the `NSNotificationCenter`.

Listing 8-3:   Overriding viewWillAppear:

```
-  (void)viewWillAppear:(BOOL)animated {


   [[NSNotificationCenter defaultCenter]
         addObserverForName:UIKeyboardWillShowNotification
         object:self.view.window queue:nil
         usingBlock:keyBoardWillShow];

   [super viewWillAppear:animated];
}
```

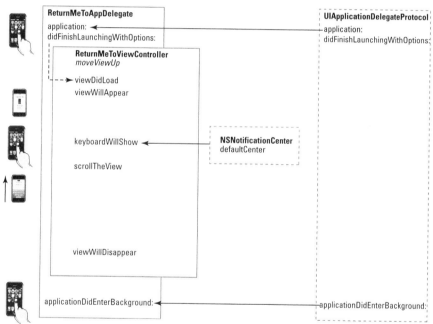

Figure 8-4:
Where to
register
for a
notification.

Here's the blow-by-blow account:

1. I send the NSNotificationCenter the defaultCenter: message, passing it the following items:

• addObserverForName: This specifies the notification you're registering for — in this case, UIKeyboardWillShow Notification.

- `object`: The particular object whose notification I am registering for — in this case, `window`.

- `queue`: The operation queue to which the block should be added. This is important if you plan on doing concurrent processing. Because I have no such plans, I pass in `nil` and the block is run on the thread I am sending the message from.

- `usingBlock`: The block to be executed when the notification is received.

So, you might be wondering ,where is this Block object? I'll get to that in "The mechanics of scrolling the view" section, later in this chapter. But before I do, I want to explain what the code needs to do. I cover that in the next section.

2. **I then invoke the superclass's (`UIViewController`) method `[super viewWillAppear:animated];`.**

This is important because there may be some things `UIViewController` needs to do on its own before the view appears.

Keeping the text field visible

It's often a challenge to implement the scrolling of your content in a view because sometimes it's difficult to determine how much scrolling you need to do — or whether you should do any scrolling at all. For example, when the keyboard appears, it may not even cover the text field you added to the view — which means that you shouldn't scroll the content view. In fact, if the text field won't be covered by the keyboard, scrolling the content might be a bad thing to do; you may end up scrolling the content out of sight.

Things are starting to get a little complicated, so I'm going to pull out the good-old, illustrative line drawing (Figure 8-4) on the theory that a picture is worth a thousand words.

The concept

The `UIKeyboardWillShowNotification` is posted by the window right after the user touches in the text field, but before the keyboard appears.

At that time, you determine whether the keyboard is going to cover the text field; if you see that it will, you set the method's instance variable `moveViewUp` to `YES`. This variable will be used after the user is done editing, to see whether the content view has been scrolled and needs to be restored.

As part of this process, you have to compute the actual amount you need to scroll. You only want to scroll the content view enough for the text field — plus a little margin — to be visible above the top of the keyboard. After computing that value, you save it in an instance variable so the method that actually scrolls the content view knows how much "scroll" to use.

That's the concept. The actual mechanics of coding this scroll business get ironed out in the next section.

The mechanics of scrolling the view

As part of the `UIKeyboardWillShowNotification` you have access to an `NSDictionary` (in the `NSNotification` object returned by the `NSNotificationCenter`) that contains, among other things, the height of the keyboard.

Dictionaries manage pairs of keys and values. Each of these key-and-value pairs is an *entry*. Each entry has two objects; one object represents the key, and a second object is that key's value. An `NSDictionary` object manages a *static array* — an array whose keys and values cannot be added to or deleted (although individual elements can be modified).

The dictionary has the four entries, as shown in Figure 8-5:

- ✔ The key `UIKeyboardFrameBeginUserInfoKey` has as its value a `CGRect` that is the start frame of the keyboard in screen coordinates before animation — that is, before it is scrolled in.

- ✔ The key `UIKeyboardFrameEndUserInfoKey` has as its value a `CGRect` that is the end frame of the keyboard in screen coordinates after animation — that is, after it is scrolled in.

- ✔ The key `UIKeyboardAnimationDurationUserInfoKey` has as its value a double that tells you how long the animation is in seconds.

- ✔ The key `UIKeyboardAnimationCurveUserInfoKey` has as its value a `UIViewAnimationCurve` constant that tells you whether the keyboard animation changes speed at the beginning or end.

You want to scroll the content view only enough so that the text field, plus a small margin, is sitting just above the top of the keyboard. Now, Figure 8-5 shows you that you can get the size of the keyboard from the `NSDictionary`. Basically, you can calculate the size of the view as well as the origin of the text field and its size. With this figure in hand, you can compute the `bottom Point`, which you can see on the right side of the figure. You can then determine the number of pixels between the bottom of the view and the calculated `bottomPoint`. And if you use that pixel value (you will in the computation), you can then determine the amount to scroll the content view.

Listing 8-4 shows the code (in bold) that implements the keyboard scrolling logic. Add it to `viewWillAppear:`. The handy steps that follow Listing 8-4 give the color commentary on what all this code is actually doing.

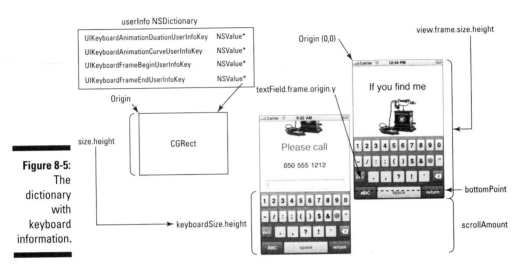

What is also interesting here is that you have put the code in a Block object that will be executed by the notification center when `UIKeyboardWill ShowNotification` is posted by the window. While it may look strange at first, it's actually pretty straightforward.

Back under OS 3.x, I defined a method in order to handle scrolling whenever the keyboard showed

```
- (void)keyboardWillShow:(NSNotification *)notif
```

Instead, in OS 4.0 I define a block, which for all practical purposes does the same thing as the method.

Blocks are objects that encapsulate code that can be executed at any time. They're essentially functions that have many of the characteristics of methods that you can pass in as arguments of methods. Blocks have the advantage of being able to access both the local variables of a method and the instance variables of the object.

Listing 8-4:   Overriding viewWillAppear:

```
- (void)viewWillAppear:(BOOL)animated {

  void (^keyBoardWillShow) (NSNotification *) =
                          ^(NSNotification * notif) {
    NSDictionary* info = [notif userInfo];
    NSValue* aValue =
        [info objectForKey:UIKeyboardFrameEndUserInfoKey];
```

```
    CGSize keyboardSize = [aValue CGRectValue].size;
    float bottomPoint = (textField.frame.origin.y +
        textField.frame.size.height + 10);
    scrollAmount = keyboardSize.height - (self.view.frame.
        size.height - bottomPoint);

    if (scrollAmount > 0)  {
      moveViewUp =YES;
      [self scrollTheView:YES];
    }
    else
      moveViewUp =NO;
    webView.userInteractionEnabled=NO;
};

[[NSNotificationCenter defaultCenter]
        addObserverForName:UIKeyboardWillShowNotification
        object:self.view.window queue:nil
        usingBlock:keyBoardWillShow];


[super viewWillAppear:animated];
}
```

Here's the blow-by-blow account:

1. **Declare a block variable by using the ^ operator, with the name of keyBoardWillShow that returns void and takes an NSNotification * as its single argument. (To know that, you have to look it up in the NSNotificationCenter class reference.)**

   ```
   void (^keyBoardWillShow) (NSNotification *)
   ```

   Just as with any other variable declaration (like int i = 1;), you follow the equal sign with its definition. You do that by using the ^ operator again to indicate the beginning of the *block literal* — the definition assigned to the block variable. The block literal includes argument names (so you also inform the compiler that the name of the argument is notif.) as well as the body (code) of the block,

   ```
    = ^(NSNotification * notif) { …
   ```

and end the block literal with the usual

   ```
    };
   ```

This is all there really is to blocks. Now I want to run through the code that actually does the scrolling. Here's what your code is going to do:

2. **Send a message to the notification object to return a reference to the dictionary that has the information.**

```
NSDictionary* info = [notif userInfo];
```

3. **Use the key to have the method extract the keyboard size for you.**

```
NSValue* aValue =
    [info objectForKey:UIKeyboardFrameEndUserInfoKey];
CGSize keyboardSize = [aValue CGRectValue].size;
```

The `NSValue` object is a simple container for a single C or Objective-C data item. It can hold any of the scalar types (variables that hold values) such as `int`, `float`, and `char`, as well as pointers, structures, and object IDs. In this case, as you can see in this code snippet — as well as in Figure 8-5 — it's a `CGRect` — a structure that contains the location (`origin`) and dimensions (`size`) of a rectangle.

`CGSize` is a structure that contains width and height values. `[aValue CGRectValue]` invokes a method that returns the `CGRect`, from which you get the size.

4. **Compute the `bottomPoint`.**

```
float bottomPoint = (textField.frame.origin.y +
                     textField.frame.size.height + 10);
```

`textField.frame.origin.y` in Figure 8-5 tells me the top-left point of the text field. To find the bottom point, I'm adding both the height of the text field and a 10-pixel margin to make it look nice. That's because the coordinate system on the iPhone starts at (0,0) in the top-left corner of the screen and increases as you go down the screen. (Forget what you learned in Algebra I, because it won't help you deal with iPhone coordinates.)

5. **Compute the amount to scroll.**

```
scrollAmount = keyboardSize.height -
        (self.view.frame.size.height - bottomPoint);
```

As you can see in Figure 8-5, subtracting the `bottomPoint` from the height of the content view gives you the amount of the content view that I want covered by the keyboard. I get the height of the content view by using the view controller's pointer to the view — the aptly named `view` pointer. The view has an instance variable, `frame`, which is a `CGRect` that has a `size`, just as the `CGRect` for the keyboard did. Subtracting that result from the keyboard height gives me the amount to scroll.

6. **Check to see whether the view should be moved up.**

```
if (scrollAmount > 0) {
```

If the scroll amount is greater than zero, I set `moveViewUp` to YES. This will be used by the methods invoked when the user is done editing the text field to see whether the content view has been scrolled and needs to be restored. If not, I set it to NO. Finally, I send the `scrollTheView :YES` message to move the view up.

Of course, if the scroll amount is not greater than zero, I set `moveViewUp` to `NO` and forget about the whole thing.

7. **I explain the following after listing 8-3, but to refresh your memory, here it is again:**

   Send the `NSNotificationCenter` the `defaultCenter:` message, passing it the following items:

   - `addObserverForName:` This specifies the notification you're registering for — in this case, `UIKeyboardWillShow Notification`.

   - `object:` The particular object whose notification I am registering for — in this case, `window`.

   - `queue:` The operation queue to which the block should be added. This is important if you plan on doing concurrent processing. Because I have no such plans, I pass in `nil` and the block is run on the thread I am sending the message from.

   - `usingBlock:` The block to be executed when the notification is received.

8. **I then invoke the superclass's (`UIViewController`) method `[super viewWillAppear:animated];`.**

   This is important because there may be some things `UIViewController` needs to do on its own before the view appears.

You may think this whole computing-and-calculating business is overkill. After all, I only have the one text field, and I *know* it's going to be covered. But there's method in the madness: I did this to introduce you to some view geometry so you can see how to compute where things are in the view and where to find the values. I could have also hard-coded the size of the keyboard, but sizes, depending on the type of the keyboard you've chosen, may be different, and keyboard sizes may also change between different releases of the iPhone OS. In addition, the keyboard size can change according to whether the device is in portrait or landscape mode (if your iPhone supports that).

It wouldn't be the first time that a quick-and-dirty method came back to haunt you, or the last for that matter.

Remember what your parents said after they made you do something that was difficult? "I'm doing it for your own good."

Fortunately, this is as hard as it gets.

Unregistering a notification

While I'm at it, I should also write the code for *un*registering for the notification. That's because I don't want the notification center to send a notification to an object that has been *freed* (that is, de-allocated). Again, a quick peek back at

Figure 8-4 shows that the framework supplies a convenient place to put that — `viewWillDisappear:` — when the user decides to switch to another view or terminate the application. (Listing 8-5 shows the code you need for unregistering the application.)

If you skip this unregistering step, you generate a runtime error if the center sends a message to a freed object.

Although the ReturnMeTo application has only one view, other applications you create could very well have other views — or you may even come back later and enhance *this* application with another view. So it's always good form to do this kind of unregistering cleanup *before* a view is either freed or is no longer visible. Here (again) you use a `defaultCenter` message, passing the following items to it:

- ✔ `removeObserver:` The object you want to send the message to. It's either going to be `self` or the object making the request, in this case the `ReturnMeToViewController`.

- ✔ `name:` This specifies the notification you're *unregistering* for — in this case, the `UIKeyboardWillShowNotification`.

- ✔ `object:` Because I'm unregistering, I use a `nil` for `object` here because that will remove all `UIKeyboardWillShowNotification` notifications (if there were more than one).

Listing 8-5 shows it all. I suggest that you add it right after `viewWillAppear:`.

Listing 8-5:   Override viewWillDisappear:

```
- (void)viewWillDisappear:(BOOL)animated {

  [[NSNotificationCenter defaultCenter] removeObserver:self
         name:UIKeyboardWillShowNotification object:nil];

  [super viewWillDisappear:animated
```

Moving the view

After you determine how much to scroll the content view for your ReturnMeTo application (see the previous section), you can then put the code in place for actually moving the view. Listing 8-6 shows the `scrollTheView:` method that must be added to the `ReturnMeToViewController.m` file. I suggest that you add it right before `didReceiveMemoryWarning`.

The steps that follow highlight the major points along the way. (As always, if you need a refresher on code writing, check out Chapter 7.)

Listing 8-6:   Adding scrollTheView:

```
- (void)scrollTheView:(BOOL)movedUp {

   [UIView beginAnimations:nil context:NULL];
   [UIView setAnimationDuration:0.3];
   CGRect rect = self.view.frame;
   if (movedUp){
     rect.origin.y -= scrollAmount;
     }
   else {
     rect.origin.y += scrollAmount;
     }
   self.view.frame = rect;
   [UIView commitAnimations];
}
```

The first step has to do with *animating* the move — programming-speak for that nice smooth sliding up and down of views you see on the iPhone. I could just simply move it, but instead I want to move it in synch with the keyboard moving up. This is called *animating the transition.* The UIView framework that you've been working with has several class methods you can use to deal with the whole animation ball of wax:

✔ Methods to indicate that the transition should be animated in the first place

✔ Methods to indicate the type of transition that should be used

✔ Methods to specify how long the transition should take

Okay, here goes, one step at a time:

1. Create an animation block.

To invoke a view's built-in animation behavior, you create an animation block and set the duration of the move.

beginAnimations:: has arguments to pass information to animation delegates. Because you're not going to be using any such delegates in the ReturnMeTo application, you should set the arguments to nil and NULL.

```
[UIView beginAnimations:nil context:NULL];
```

nil is used when there is a null pointer to an object — begin Animations::, for example.

NULL is used when there is a null pointer to anything else.

As for animation duration, I set that to 0.3 seconds, which matches the keyboard's animation.

```
[UIView setAnimationDuration:0.3];
```

2. **Get (access) the view's frame.**

   I did the very same thing in the `keyboardWillShow:` method back in Listing 8-3.

   ```
   CGRect rect = self.view.frame;
   ```

3. **If the view should be moved up, subtract the keyboard height from the origin.**

   The `CGRect` also contains the view's `origin` in *x, y* coordinates, with the upper-left part of the screen being 0,0.

   ```
   if (movedUp) {
        rect.origin.y -= scrollAmount;
   ```

4. **If the view shouldn't be moved up, restore it by adding the keyboard height back to the origin.**

   ```
   else {
        rect.origin.y += scrollAmount;
   ```

   If I move the content view up when the keyboard appears, then I must also restore the view to its original position when the keyboard disappears. This code allows me to send the `scrollTheView:` message with NO, which will scroll the view down.

5. **Assign the new frame to the view.**

   ```
   self.view.frame = rect;
   ```

6. **Tell the view that you're all done with setting the animation parameters, and it should start the animation.**

   ```
   [UIView commitAnimations];
   ```

Changing the frame rectangle automatically redisplays the view! You don't have to lift a finger! Remember, though, that because you set the new frame inside an animation block, the view doesn't instantly move to the new position, but is instead animated over time (0.3 seconds in this case) to the new frame position.

Updating the interface

Time for a bit of cleanup. So far in this chapter, you've made quite a few changes to the original code, so it's probably time to bring the interface up to speed on what you've been mucking about with. Listing 8-7 shows what you need to add to the interface to let it know what's been happening on its watch. (Note that the changes are in bold.) You can see right off the bat that you need to declare the two new instance variables, `moveViewUp` and `scrollAmount` — as well as the new method `scrollTheView:` — in the interface.

As to where to put them, the two instance variables go in the `ReturnMeTo ViewController.h` file, right after the two outlets we added in the last chapter. The new method declaration should follow the property declarations we added in the last chapter.

Listing 8-7:   Adding Changes to the Interface

```
@interface ReturnMeToViewController : UIViewController {

  IBOutlet UITextField *textField;
  IBOutlet UILabel    *label;
         BOOL          moveViewUp;
         CGFloat       scrollAmount;
}
@property (nonatomic, retain) UITextField *textField;
@property (nonatomic, retain) UILabel *label;

- (void)scrollTheView:(BOOL)movedUp;
@end
```

To inform the interface of what you've been up to, follow these steps:

1. **Add the `moveViewUp` and `scrollAmount` instance variables.**

   `moveViewUp` will be used to determine whether the content view needs to be — or has been — scrolled. `scrollAmount` is the amount the content view needs to be scrolled.

2. **Declare the `scrollTheView:` method.**

Lowering the view when all is said and done

The way you've set up the ReturnMeTo application is that, after the user is done entering the necessary text into the text field — in this case, a telephone number — the keyboard is supposed to disappear and the view is meant to move back down to its original position. (If you remember, you added the capability to do this in Listing 8-6, and more specifically in Step 4 in the section "Moving the view.")

This is a two-step process. First, the `textFieldShouldReturn:` message of the `UITextViewDelegate` is sent when the user taps the Return key on the keyboard, giving you the opportunity to add any functionality needed when the user is done entering text. As you can see in Figure 8-6, because the user is done editing, this is the perfect place to add code that lowers the keyboard. You do that by adding the code shown in Listing 8-8 to the `ReturnMeToViewController.m` file, right after the `dealloc` method, but before the `@end` statement.

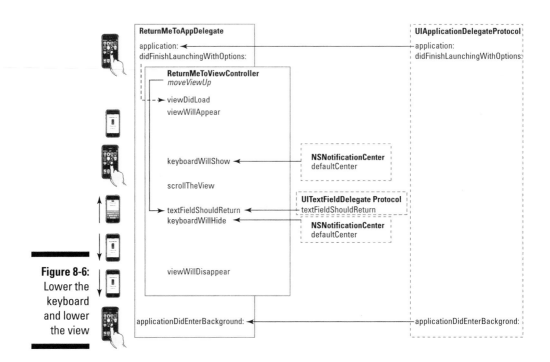

Figure 8-6:
Lower the
keyboard
and lower
the view

Listing 8-8:   Implementing textFieldShouldReturn:

```
-(BOOL)textFieldShouldReturn:(UITextField *)
                                    theTextField {

    [theTextField resignFirstResponder];

    return YES;
}
```

The method textFieldShouldReturn: is passed the current text field that
is being edited — namely, theTextField.

1. **Send [theTextField resignFirstResponder] message.**

   If you were wondering how to dismiss the keyboard, this does the trick:

   ```
   [theTextField resignFirstResponder];
   ```

2. **Return YES.**

   This tells the text field to implement its default behavior for the Return
   key.

Managing the keyboard

When the user taps a view, that view becomes the *first responder* — the first object in the responder chain given the opportunity to respond to an event. (For more on first responders and responder chains and events, see Chapter 6.) If the view is a text field (or any other object that has editable text), an editing session starts, and the keyboard is displayed automatically — you don't have to lift a finger to make the keyboard appear.

But just because the keyboard is *displayed* automatically, that doesn't mean the keyboard will be *dismissed* automatically. In fact, it's your responsibility to dismiss the keyboard at the appropriate time, when the user taps the Return or Done button on the keyboard or, in this case, when the user touches in the view.

To dismiss the keyboard, you send the resignFirstResponder message to the text field — the initial first responder. When the text field resigns, it is no longer the first responder — just like Nixon was no longer president of the United States after he resigned.

This may sound a bit convoluted, but it's the only way to dismiss the keyboard. You can't send a message to the keyboard directly — as in, "Hey, you're fired. Pack your things and go!" You can only make the keyboard disappear by having it resign its first-responder status of the text field.

While this lowers the keyboard, it doesn't restore the view to its original position. While you could (and people often do) do this in textField ShouldReturn: a better way to do this is by again taking advantage of the notification system. In Listing 8-9, I have added (in bold) more code to viewWillAppear: to register for the UIKeyboardWillHideNotification and to specify the Block object to be executed when that happens. You can see where that will happen back in Figure 8-6.

Listing 8-9:   Adding to viewWillAppear:

```
- (void)viewWillAppear:(BOOL)animated {

  void (^keyBoardWillShow) (NSNotification *) =
                          ^(NSNotification * notif) {
    NSDictionary* info = [notif userInfo];
    NSValue* aValue =
        [info objectForKey:UIKeyboardFrameEndUserInfoKey];
    CGSize keyboardSize = [aValue CGRectValue].size;
    float bottomPoint = (textField.frame.origin.y +
          textField.frame.size.height + 10);
    scrollAmount = keyboardSize.height - (self.view.frame.
        size.height - bottomPoint);

    if (scrollAmount > 0)  {
```

(continued)

Listing 8-9 *(continued)*

```
      moveViewUp =YES;
      [self scrollTheView:YES];
   }
   else
      moveViewUp =NO;
};

[[NSNotificationCenter defaultCenter]
      addObserverForName:UIKeyboardWillShowNotification
      object:self.view.window queue:nil
      usingBlock:keyBoardWillShow];

void (^keyBoardWillHide) (NSNotification *)=
                              ^(NSNotification * notif) {
   if (moveViewUp) [self scrollTheView:NO];
};
[[NSNotificationCenter defaultCenter]
      addObserverForName:UIKeyboardWillHideNotification
      object:self.view.window queue:nil
      usingBlock:keyBoardWillHide];

[super viewWillAppear:animated];
}
```

I've done the following here:

1. **I add the code that makes up the Block object.**

2. **Within that block, I check to see if the view has been scrolled up (yes, this is one of those situations), and send the `scrollTheView` message with the argument of `NO` to restore the view to its original position:**

   ```
   if (moveViewUp) [self scrollTheView:NO];
   ```

   `moveViewUp` lets me know that the view has been scrolled and it needs to be restored.

3. **I send the `NSNotificationCenter` the `defaultCenter:` message, passing it the following items:**

   - `addObserverForName:` This specifies the notification you're registering for — in this case, `UIKeyboardWillHideNotification`.

   - `object:` The particular object whose notification I am registering for — in this case, `window`.

   - `queue:` The operation queue to which the block should be added. This is important if you plan on doing concurrent processing. Because I have no such plans, I pass in `nil` and the block is run on the thread I am sending the message from.

- • usingBlock: The block to be executed when the notification is received. In this case, the expected block takes only one argument, (NSNotification *) — the notification we want — and returns a void. (To know that, you'd have to look it up in the NSNotificationCenter class reference.)

4. Correspondingly, I add the code in bold in Listing 8-10 to unregister for the notification.

Listing 8-10: Update viewWillDisappear:

```
- (void)viewWillDisappear:(BOOL)animated {

  [[NSNotificationCenter defaultCenter] removeObserver:self
        name:UIKeyboardWillShowNotification object:nil];
  [[NSNotificationCenter defaultCenter] removeObserver:self
        name:UIKeyboardWillHideNotification object:nil];

  [super viewWillDisappear:animated];
}
```

Now the keyboard has been dismissed and the view restored to its original position. You've completed the five-step plan that puts in place a system for efficiently scrolling your content view up and down as needed. Now all you have to do is tie up a few loose (code) strings and you're set — except for the compiling and testing, of course.

Polishing the Chrome and Adding the Vinyl Pinstriping

It should come as no surprise to you that the star of the last section — the aptly named textFieldShouldReturn: method — is a method within a particular protocol, just like the application:didFinishLaunchingWith Options: and applicationDidEnterBackground: methods were also methods within a protocol. In this case, the textFieldShouldReturn: method is a member of the UITextFieldDelegate protocol — the protocol that sets the rules for messages sent to a text field delegate as part of the editing sequence.

Now, protocols simply declare methods that can be implemented by any class. In response to certain events, the framework checks to see whether there's a delegate that implements a certain method — and if there is, it will invoke that method. That means if I have my ReturnMeToViewController class adopt the UITextFieldDelegate protocol and then implement the textField ShouldReturn: method, the UITextFieldDelegate protocol kicks into action, and the textFieldShouldReturn: method is invoked automatically.

Sounds great! But you do have to jump through a few hoops to have your `ReturnMeToViewController` class adopt the `UITextFieldDelegate` protocol. Here's the bird's-eye view in two easy steps:

1. **Signal to the compiler that the `TextFieldDelegate` protocol has been adopted by the delegate — in this case, `ReturnMeToViewController`.**

2. **Connect the `ReturnMeToViewController` to the `textField` to let text Field know in no uncertain terms that `ReturnMeToViewController` is its delegate.**

That's the bird's-eye view. The next few sections show what the process looks like down on the ground.

Adopting a protocol

Adopting a protocol is a pretty straightforward process. You'll be working with the header file for your class — `ReturnMeToViewController.h`, in this case. The idea here is to update the `@interface` declaration so the left hand knows what the right hand is doing.

The change is in bold in Listing 8-11.

Listing 8-11:   Adding a Protocol Declaration to the Interface

```
@interface ReturnMeToViewController : UIViewController
                                      <UITextFieldDelegate>
           {

  IBOutlet UITextField *textField;
  IBOutlet UILabel     *label;
           BOOL        moveViewUp;
           CGFloat     scrollAmount;
}

@property (nonatomic, retain) UITextField *textField;
@property (nonatomic, retain) UILabel *label;

- (void)scrollTheView:BOOL) movedUp;
@end
```

Here's the blow-by-blow account:

1. **Add `<UITextFieldDelegate>`.**

   Listing a protocol within angle brackets after the superclass name — `UIViewController`, in this case, the Big Papa class of your `ReturnMeToViewController` class — specifies that your class has adopted the `UITextFieldDelegate` protocol.

Classes can adopt several protocols. To add more than one protocol, you just put them all in the angle brackets, separated by commas.

2. Save the file. You're done.

Not too many blows there, as you can see. On to the next section.

Connecting things up with Interface Builder

In Chapter 7, you work a bit with Interface Builder to connect outlet values to their corresponding view elements. In this section, you have to trundle out Interface Builder again in order to connect `ReturnMeToViewController` with the appropriate text field so that the connections are set up for you at runtime. Here's how the process works:

1. **Back in Xcode, open the Resources folder for your ReturnMeTo project in the Groups & Files list on the left and then double-click the `ReturnMeToViewController.xib` file.**

   Interface Builder opens onscreen, displaying the main nib window and the View window open for inspection.

2. **Right-click the File's Owner icon in the main nib window.**

   A dialog appears, listing the various connections for File's Owner.

3. **Select the text field and right-click it.**

   A pop-up menu appears, listing the text field's connections. Under the first section, Outlets (expand it if it isn't expanded), you see Delegate.

4. **Drag from the little circle next to the text field's Delegate item listed under Outlets in the pop-up menu to the File's Owner icon in the main nib window, as shown in Figure 8-7.**

   The File's Owner (`ReturnMeToViewController`) is now the delegate for the text field. (Want proof? Check out the pop-up menu for the text field in Figure 8-8, which lists File's Owner as the delegate under Outlets.)

This completes what you need to do to implement scrolling, but there are still a couple more things I recommend that you do to make everything shiny and bright. My detailing list includes the following:

- Adding a Clear button to make it more convenient for the user
- Adding a feature for the user so that touching anywhere in the view does the same thing as tapping Return on the keyboard
- Saving the phone number the user enters for future reference and then displaying it in the label

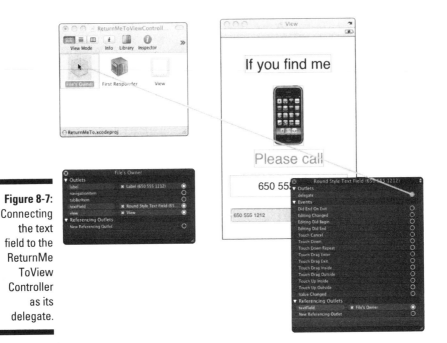

Figure 8-7:
Connecting
the text
field to the
ReturnMe
ToView
Controller
as its
delegate.

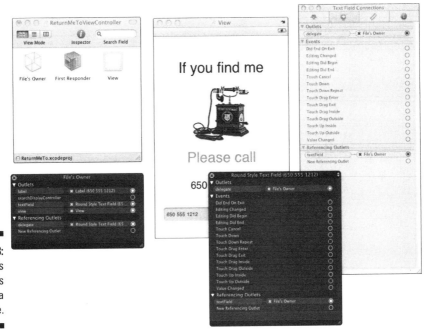

Figure 8-8:
The File's
Owner is
now a
delegate.

Because adding these extras will enable you to delve even deeper into the mechanics of Objective-C, I strongly recommend that you finish up with these details.

Adding a Clear button

I mention earlier in the chapter (and even give visual proof in Figure 8-3) that the viewDidLoad method — one of the methods generated (albeit commented out) by Xcode when you chose the View-Based Application template in Chapter 4 — is the place to do your view initialization tasks, including adding functionality not specified in the nib file. As you may have guessed, this is the perfect place to add a Clear button to the text field, which I do in Listing 8-10.

Find viewDidLoad in the ReturnMeToViewController.m file by pressing ⌘+F and entering viewDidLoad in the Find field. Uncomment it out and add the code in Listing 8-12.

Listing 8-12:   Overriding viewDidLoad

```
- (void)viewDidLoad {

  textField.clearButtonMode =
                          UITextFieldViewModeWhileEditing;
  [super viewDidLoad];
}
```

clearButtonMode controls when the standard Clear button appears in the text field. In this particular mode, the clear button is shown only while editing, although there are other choices including never and always.

That was easy!

Saving the phone number for future reference

To save the phone number the user entered, you have to add a new method — the updateCallNumber method — to ReturnMeToView Controller.m. Place it after textFieldShouldReturn:. This method — shown in Listing 8-13 — simply saves the text and assigns the text to the label. You use it here to keep track of the number the user has entered.

Listing 8-13: Adding updateCallNumber

```
- (void)updateCallNumber {
        self.callNumber = textField.text;
        label.text = self.callNumber;
}
```

Here's what's up with Listing 8-13:

1. **Store the text of the text field in the `callNumber` instance variable.**

    ```
    self.callNumber = textField.text;
    ```

    That's why you added an `IBOutlet` for the text field back in Chapter 7. It enables you to get the text the user enters.

2. **Set the text of the label to the value of the `callNumber` instance variable.**

    ```
    label.text = self.callNumber;
    ```

    And that's why you needed an `IBOutlet` for the label — to be able to update it with the new number.

`updateCallNumber` saves the number data. But even though it's implemented in the view controller, `updateCallNumber` is really, in part, a model function. In this sample program, though, there is no separate model. In reality, there isn't that much for the model to do, except save and return the saved phone number. Instead of adding complexity to the ReturnMeTo application, I decided that it was far easier and less complex to simply put the code in the view controller. (Of course, I *do* use a model when I develop the iPhone Travel411 application; I discuss model design and use when I talk about the MobileTravel411 and iPhoneTravel411 designs in Chapter 13 and implement a model in Chapter 16.)

At this point, you also need to add the `callNumber` instance variable and the corresponding `@property` and `@synthesize` statements. Read on to find out how.

1. **In the `ReturnMeToViewController.h` file, add the following instance variable:**

    ```
    NSString    *callNumber;
    ```

2. **Then add the property declaration (see Listing 8-14):**

    ```
    @property (nonatomic, retain) NSString *callNumber;
    ```

3. **In the `ReturnMeToViewController.m` file, add the `@synthesize` statement after the `@synthesize` label statement you added in Chapter 7, as you can see in Listing 8-15.**

Listing 8-14: Add callNumber to the Interface

```
@interface ReturnMeToViewController : UIViewController
                                      <UITextFieldDelegate> {

        IBOutlet UITextField *textField;
        IBOutlet UILabel     *label;
                 BOOL         moveViewUp;
                 CGFloat      scrollAmount;
                 NSString    *callNumber;

}

@property (nonatomic, retain) UITextField *textField;
@property (nonatomic, retain) UILabel *label;
@property (nonatomic, retain) NSString *callNumber;
```

Listing 8-15: Synthesizing the accessors

```
#import "ReturnMeToViewController.h"
#import "ReturnMeToAppDelegate.h"

@implementation ReturnMeToViewController

@synthesize textField;
@synthesize label;
@synthesize callNumber;
```

You added the updateCallNumber method. That's great — but to actually save the number and display it in the label, someone has to send the updateCallNumber message when the user is actually done entering the number. (Duh.) One of the places you need to do that is in the textField ShouldReturn: method. At that point in the process, the user has finished entering the phone number and has tapped Return on the keyboard. In Figure 8-9, I added the updateCallNumber method and show textField ShouldReturn: invoking it. Note that I placed the method off to the side in the diagram, in an area labeled *Model*. This is to let you know that while the method is being implemented in the ReturnMeToViewController class, it is (conceptually, at least) a model method. Listing 8-16 shows the modifications I made to textFieldShouldReturn: in bold.

Listing 8-16: Modifying textFieldShouldReturn:

```
- (BOOL) textFieldShouldReturn: (UITextField *) theTextField {

    [theTextField resignFirstResponder];
    [self updateCallNumber];

    return YES;
}
```

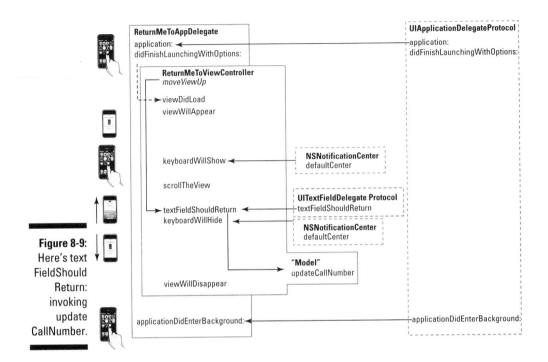

Figure 8-9:
Here's text
FieldShould
Return:
invoking
update
CallNumber.

Adding a method here means that I need to add the method declaration to
the interface as well. Listing 8-17 shows how to code that addition.

Listing 8-17: Adding Changes to the Interface

```
@interface ReturnMeToViewController : UIViewController
          <UITextFieldDelegate> {

   IBOutlet UITextField *textField;
   IBOutlet UILabel     *label;
            BOOL         moveViewUp;
            CGFloat      scrollAmount;
            NSString    *callNumber;
}

@property (nonatomic, retain) UITextField *textField;
@property (nonatomic, retain) UILabel *label;
@property (nonatomic, retain) NSString *callNumber;


- (void)scrollTheView:(BOOL)movedUp;
- (void)updateCallNumber;

@end
```

Dismissing the keyboard when the user touches in the view

I want the keyboard to disappear when one of two things happens:

- ✔ The user taps the Return button on the keyboard.
- ✔ The user touches anywhere else in the view.

This lets you know that the user is done entering text and doesn't need the keyboard any longer.

The Multi-Touch interface is one of the things that makes the iPhone so appealing to users because direct manipulation makes people feel more in control. That means you want to implement direct manipulation (or touches) not only when it might be useful for a user, but also when the user might expect it.

I'm going to start your trip into finding out how to process touches on the view, in a control, and as gestures by showing you how to make the keyboard disappear when the user touches in the view.

I already implemented the first requirement in `textFieldShouldReturn:`. Figure 8-10 shows the method `touchesBegan::`. I'm overriding a method of the `ReturnMeToViewController`'s superclass, `UIResponder`, from which the view controller is derived. The `touchesBegan::` message is sent when one or more fingers touches down in a view. The implementation of that is shown in Listing 8-18. Notice `touchesBegan::` also references `moveViewUp` to determine whether it should send the `scrollTheView:` message to restore the content view.

Listing 8-18:   Override touchesBegan::

```
- (void)touchesBegan:(NSSet *)touches withEvent:
                                        (UIEvent *)event {
    if( textField.editing) {
      [textField resignFirstResponder];
      [self updateCallNumber];
    }
    [super touchesBegan:touches withEvent:event];
}
```

The following steps break down Listing 8-18 into its constituent parts:

1. See whether the text field is currently being edited.

```
   if( textField.editing) {
```

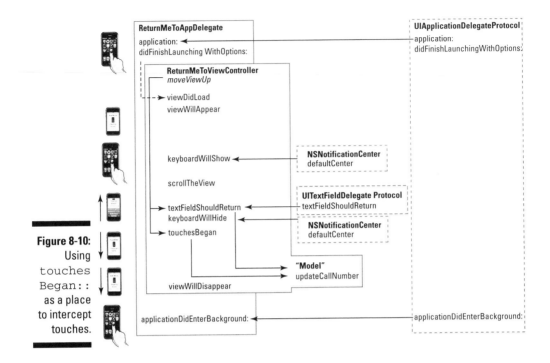

The user may touch the view any time, not just when he or she is done entering a phone number. Every time the user touches in the view, the `touchesBegan::` message is sent, and I shouldn't go through the code to resign as first responder and update the call number if the user hasn't really finished the editing job yet.

`textField.editing` is a Boolean value that indicates whether the text field is currently in edit mode. I only process the touch then — when I'm in editing mode.

2. **Send the `resignFirstResponder:` message.**

   That will cause the keyboard to disappear.

3. **Send the `updateCallNumber` message.**

   This method updates the text in the label with the text that the user typed, and then saves the text in an instance variable.

4. **Send the `[super touchesBegan:touches withEvent:event];` message.**

   This passes the event to the superclass if it needs to do anything more.

Because you used an accessor method to assign a value to `callNumber` — remember the `@property` and `@synthesize` statements you added? — it was sent a `retain` message. That makes you responsible for releasing it (freeing up the memory), just as you were for the outlets in Chapter 7. Listing 8-19 shows you how.

Listing 8-19:   Releasing the callNumber

```
- (void)dealloc {

   [textField release];
   [label release];
   [callNumber release];

   [super dealloc];
}
```

Finding Your Way Around the Code

When you look in Listing 8-18 (it's on my Web site), you can see several #pragma statements. For example, this one:

```
#pragma mark - UIViewController methods
```

Any statement that begins with #pragma is actually a compiler directive — meaning that it has nothing to do with code; it just passes information to the compiler or, in this case, code editor. It tells Xcode's editor to put a "heading" in the pop-up menu at the top of the Editor Pane that stores a running list of methods and functions used in the project. You get this pop-up list by selecting the up and down arrows highlighted in Figure 8-11. The field to the right displays the last method you were in, in this case viewDidLoad. Choose a method or function from the pop-up menu, and you're brought to the implementation of that method in the code.

Some of your classes, especially some of your controller classes, are likely to have a lot of code. If you make it a habit to use this gem of a pop-up menu, you'll find it much easier to find things.

When You're Done

If you head over to my Web site at www.nealgoldstein.com, you'll find Listings 8-20 and 8-21, which show the changes I've made to the original code in Listings 8-1 and 8-2 — the whole kit and caboodle.

When you compile and run this code in Xcode, using the handy Build and Run button on the toolbar, you end up with the obediently scrolling iPhone view highlighted back in Figure 8-1 on the right. You still have some work to do, though. (No rest for the weary.) For starters, you need to set up a way to save the data entered by your user — and when you're done with that, I have a few surprises for you.

More on that in the next chapter.

Up and Down Arrows

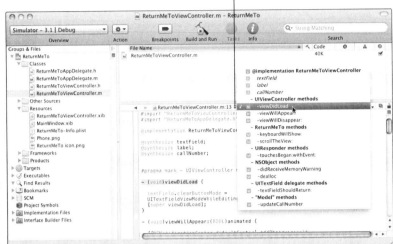

Figure 8-11:
Finding the
pop-up list.

Chapter 9

Saving Data and Creating a Secret Button

In This Chapter

▷ Setting and accessing user preferences

▷ Saving data in user preferences

▷ Disabling and enabling text fields

▷ Responding to user touches

*I*n putting together a great iPhone application, a big part of the whole process involves getting your application to work well from the user interface perspective. Your potential user should be able to scroll the keyboard, work with text fields, enter stuff, delete stuff, admire your fashion sense when it comes to images and background color, and generally have a grand-old time exploring the corners of your app.

Interfaces are important — so important, in fact, that most of the chapters so far in this part deal explicitly with how to set up a user-friendly interface — but interfaces are not the only things in the iPhone app universe. For your application to function as an application, it has to do application-like stuff. For example, it has to be able to save data entered by a user for the next time he or she fires up that app. In this chapter, I tackle how to get your app to save data entered by the user. Again, I'm going to trot out my ReturnMeTo application as a means of imparting this little lesson in data saving.

Wait! That's not all! It occurred to me while I was showing people my handy ReturnMeTo application that whoever found my iPhone could accidently enter a number in the text field — corrupting the crime scene, as it were — and then wouldn't know where to call me. So I decided that after the user saved a number for the first time, I would make it a little more of a challenge to write over that data. If there's a saved number, I'm going to disable the text field — and require the user to know where to tap in the content view to enable it.

Saving user entry and then controlling the editing process are two good skills to master when you're developing iPhone applications. By the end of this chapter, you should be an old hand at both skills.

Saving User-Entry Data

The iPhone is, first and foremost, a *phone;* as such, you would expect it to be able to deal with numbers — phone numbers, especially. So it's quite reasonable for the user to expect the ReturnMeTo application to save phone numbers entered into the user-entry text field. (And quite frankly, what would be the point of the application if it didn't?)

The iPhone gives you three ways to save a phone number:

- ✔ **Save the number in a file.** A perfectly respectable option, which I discuss in greater detail in Chapter 15.

- ✔ **Save the number in a database.** iPhone has a built-in SQL database that's efficient at storing and retrieving large amounts of (structured) data.

- ✔ **Save the number as an application preference.** The iPhone provides support for user preferences — allowing users to customize applications or keep track of configuration settings from launch to launch. (Hmm, the phone number to call if you lose your iPhone comes to mind here.)

So many choices. But I'm definitely going with Door #3. Let me explain why.

Preferences

Most people these days have spent enough time around computers that they know what I mean when I throw the term "preferences" around. On your desktop, for example, you can set preferences at the system level for things like security, screen savers, printing, and file sharing . . . just to name a few. But keep in mind that preferences aren't just a system-level thing; you can just as easily set preferences at the application level. You could, for example, set all sorts of preferences in Xcode — not to mention all those preferences in your browser and word-processing programs.

The latter are application-specific settings used to configure the behavior or appearance of an application. On the iPhone, application preferences are supported as well, but instead of having to (re-) create a user interface for each separate application, the iPhone displays all application-level preferences through the system-supplied Settings application. (Its icon looks like a bunch of gears on your iPhone's home screen.) Okay, you don't have to forgo creating a separate settings feature in your application, but keep this in mind: Whatever separate settings feature you come up with has to function within the framework of iPhone's Settings application; in effect, the Settings application makes you color within the lines.

What (guide)lines does the iPhone impose? Here's a short summary:

- ✔ **If you have preference values that are typically configured once and then rarely changed:** Leave the task of setting preferences to the system Settings application. On an iPhone, this would apply to things like enabling/disabling Wi-Fi access, setting wallpaper displays, setting up mail accounts, and any other preference you would set and then leave in place for a while.

- ✔ **If you have preference values that the user might want to change regularly:** In this situation, you should consider having users set the options themselves in your application.

The iPhone's weather app is a good example: Suppose I have this thing for Dubrovnik — where it happens to be 48 degrees F as I'm writing this — and I'd like to add it to my list of preferred cities that I want the weather app to keep tabs on. To load Dubrovnik into the weather app, all I have to do is tap the info button at the bottom of the screen; the view will flip around, and I can add it to my list of cities. That's a lot easier than going back to the home screen, launching the Settings application, adding the new city, and then launching the Weather application again.

The reason I'm leading you down this path is not because I'm about to show you how to use the Settings application to set user preferences — that actually comes in Chapter 15, in due time — but because the iPhone has a built-in, easy-to-use class that lets you read and set user preferences — NSUserDefaults. It's even used by the Settings application itself, which has graciously consented to let us peons use it as well — and I'm going to show you how to put that power to work so that your application can both read and set user preferences.

The NSUserDefaults class

You use NSUserDefaults to read and store preference data to a defaults data base, using a key value, just as you access keyed data from an NSDictionary. (For more on key-value pairs in general and NSDictionary in particular, see Chapter 8.) The difference here is that NSUserDefaults data is stored in the file system rather than in an object in memory — objects, after all, go away when the application terminates.

By the way, don't ask me why they stick Defaults in the name rather than something to do with preferences — fewer letters, maybe — but that's the way it is. Just don't let their naming idiosyncrasies confuse you.

Storing the data in the file system instead of in memory gives me an easy way to store application-specific information. With the help of NSUserDefaults, you can easily store the state the user was in when he or she quit the application — or store something simple like a phone number — which just so happens to be precisely what I need for the ReturnMeTo application.

Saving data by using NSUserDefaults

Enough background information; it's time to actually save some data to NSUserDefaults.

The first thing you need to decide is where in your application you plan on loading and then saving your data. As Figure 9-1 makes clear, the obvious places to do that are in application:didFinishLaunchingWithOpti ons: and applicationDidEnterBackground: — the very same methods I use in Chapter 8 to perform initialization and all the things you need to do to put your application on hold while it is in background.

As I explain in Chapter 6, the reason you want to make sure that you save the phone number in applicationDidEnterBackground: is because there's a chance that your application might be purged in background if the system runs low on memory, and if it is, you'll get no notification whatsoever — I mean nothing — that this is about to happen.

I'm going to start this data-saving business by showing you how to save the phone number. After all, being able to read the phone number from the defaults database isn't all that useful if there's nothing there to read.

(It turns out that the first time you start the application there's nothing to load, but I show you how to deal with it shortly.)

Setting it up

Because I'm going to be doing all the work in the ReturnMeToAppDelegate, I'm going to first declare an instance variable that will hold the number that needs to be saved.

1. **Add a new instance variable called savedNumber and declare @property in the ReturnMeToAppDelegate.h file.**

   Property declarations tell the compiler that there are going to be accessors for an instance variable, making it available to other objects. Because the ReturnMeToViewController object is going to have to be able to read and write the savedNumber value, the accessors have to be there.

   This is shown in Listing 9-1. (Again, the new stuff is bold.)

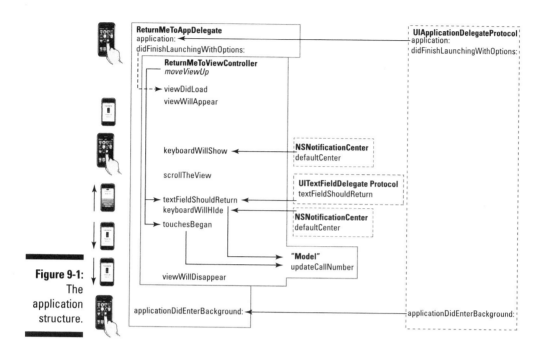

Figure 9-1:
The
application
structure.

2. **Add the `@synthesize` statement to the `ReturnMeToAppDelegate.m` file to let the compiler know that you want it to do all the work and create the accessors for you.**

This is shown in Listing 9-2. (You guessed it — new is bold.)

Listing 9-1: Adding the Instance Variable to the Interface

```
@class ReturnMeToViewController;

@interface ReturnMeToAppDelegate : NSObject
                         <UIApplicationDelegate> {

UIWindow                 *window;
ReturnMeToViewController *viewController;
NSString                 *savedNumber ;
}


@property (nonatomic, retain) IBOutlet UIWindow *window;
@property (nonatomic, retain) IBOutlet
         ReturnMeToViewController *viewController;
@property (nonatomic, retain) NSString *savedNumber;


@end
```

Listing 9-2:   Adding the synthesize to the Implementation

```
@implementation ReturnMeToAppDelegate

@synthesize window;
@synthesize viewController;
@synthesize savedNumber;
```

After the instance variable is there, to make it feel useful, I'm going to have to update it with the phone number. If you recall, back in Chapter 8, you added the `updateCallNumber` method to keep track of the number the user entered. Well, there was method to the madness because I now have a handy place to update the `ReturnMeToAppDelegate`'s `savedNumber` whenever the user changes it.

When the user enters a new phone number, you need to update the `saved Number` instance variable (which in turn will be saved in the `application DidEnterBackground:` method). You do this by adding some code to the `updateCallNumber` method, which you had earlier placed in the `ReturnMeToViewController.m` file. You can see the result in Listing 9-3.

Listing 9-3:   Updating updateCallNumber

```
- (void)updateCallNumber {

    self.callNumber = textField.text;
    label.text = self.callNumber;

    ReturnMeToAppDelegate *appDelegate =
                [[UIApplication sharedApplication] delegate];
    appDelegate.savedNumber = self.callNumber;
}
```

Now, here's the big question: If the `savedNumber` is in the `ReturnMeToAppDelegate` object, how do you get to it? Here's how:

1. **Get a reference to the `ReturnMeToAppDelegate`.**

   It turns out that this is done so often that there's a really easy way to do it. All you do is send a message to the `UIApplication` and ask for the delegate

   ```
   [[UIApplication  sharedApplication] delegate]
   ```

   This returns the delegate object — in this case, `ReturnMeToAppDelegate` — which I assign to local variable `appDelegate`.

Now you can access the application delegate and assign the number the user entered (`callNumber`) to the `ReturnMeToAppDelegate`'s instance variable (`savedNumber`), which will then be saved in `applicationWill`

Terminate:. The changes the user made are kept safe and sound, ready to appear again the next time the application is launched.

Saving the phone number

It's pretty much all downhill from here. When the user accepts the phone call, SMS, or calendar notification or presses the Home button or the system launches another application, your application then moves into background, as shown back in Chapter 6, and you'll save the number in the standard UserDefaults database. As Figure 9-1 makes abundantly clear, the place to do so is in the applicationDidEnterBackground: method. The method is already there for you in the template, so all you have to do is add the code in bold shown in Listing 9-4.

Listing 9-4:   Overriding the applicationDidEnterBackground:

```
- (void)applicationDidEnterBackground:
                            UIApplication *)application {

  [[NSUserDefaults standardUserDefaults]
      setObject:savedNumber forKey:kNumberLocationKey];
}
```

Although it's just a single statement, it's pretty complex, so let me take you through it.

It's really easy to both access and update a preference — as long as you have NSUserDefaults by your side. The trick here is to use the NSUser Defaults class to read and update whatever the user ends up entering as the phone number. NSUserDefaults is implemented as a *singleton,* meaning that there's only one instance of NSUserDefaults running in your application. To get access to that one instance, I invoke the class method standardUserDefaults:

```
[NSUserDefaults   standardUserDefaults]
```

standardUserDefaults returns back the NSUserDefaults object. As soon as I have access to the standard user defaults, I can store data there and then get it back when I need it. To store data, I simply give it a key and tell it to save the data by using that key.

The way I tell it to save something is by using the setObject;forKey: method. In case your knowledge of Objective-C is a little rusty (or not there at all), that's the way any message that has two arguments is referred to.

The first argument, setObject:, is the object I want NSUserDefaults to save. This object must be NSData, NSString, NSNumber, NSDate, NSArray, or NSDictionary. In our case, savedData is an NSString, so we're in good shape.

The second argument is forKey:. To get the data back (and for NSUserDefaults to know where to save it), I have to be able to identify it to NSUserDefaults. I can, after all, have a number of preferences stored in the NSUserDefaults data base, and the key tells NSUserDefaults which one I'm interested in. The particular key I'm using is kNumberLocationKey, which I am going to add to the ReturnMeToAppDelegate.m file, right after the last #import statement, as you can see in Listing 9-5.

If you plan to allow devices not running iOS4 to run your application, you'll also have to put this same code into the applicationWillTerminate: method as well.

Listing 9-5:   Adding the Key to ReturnMeToAppDelegate.m

```
#import "ReturnMeToAppDelegate.h"
#import "ReturnMeToViewController.h"

NSString *kNumberLocationKey = @"NumberLocation";
```

When I save the phone number, as I did in Listing 9-4, I tell NSUserDefaults to save it with a key of kNumberLocationKey. The key needs to be a string (NSString) (and so do keys in NSDictionary, by the way, which this is very similar to). Then when I want the data back (which I show you in a second), I just ask for it with that key.

Loading the preference entry to get the data

To get the phone number back — now that it's out there when the application is launched — all I need to do is ask for it with the key.

Whoops, what about the very first time the application is launched, when there is no data out there yet? Let me show you how to take care of that. In the application:didFinishLaunchingWithOptions: method, you're going to need to do two things:

- ✔ First check to see whether the preference entry exists. If one doesn't exist, you have to create one.

- ✔ If the preference does exist, you're home free and you can read it.

Listing 9-6 shows the code necessary for accomplishing these two tasks. application:didFinishLaunchingWithOptions: is in the ReturnMeToAppDelegate.m file — you have to add the code in bold.

I am working now with the ReturnMeToAppDelegate interface and implementation.

Listing 9-6: Updating application:didFinishLaunchingWithOptions:

```
- (BOOL)application:(UIApplication *)application
    didFinishLaunchingWithOptions:
                         (NSDictionary *)launchOptions {

  self.savedNumber =
            [[NSUserDefaults standardUserDefaults]
                   objectForKey:kNumberLocationKey];
  if (savedNumber == nil) {
    savedNumber = @"650 555 1212";
    NSDictionary *savedNumberDict = [NSDictionary
           dictionaryWithObject:savedNumber
           forKey:kNumberLocationKey];
    [[NSUserDefaults standardUserDefaults]
                   registerDefaults:savedNumberDict];
  }

  [window addSubview:viewController.view];
  [window makeKeyAndVisible];
}
```

Here's the blow-by-blow:

1. **Access the preference entry and save it in an instance variable.**

   ```
   self.savedNumber =
               [[NSUserDefaults standardUserDefaults]
                   objectForKey:kNumberLocationKey];
   ```

   We did something similar when we saved the number back in Listing 9-4. Here, we're using a NSUserDefaults object as well, but this time we send the setObject;forKey: message. Reading is a little easier than saving, because you have to give it only one argument: the key that you used to save the data in the first place. objectForKey: will return an Objective-C object like an NSString, NSDate, NSNumber, or any of the other types I mention earlier. (Again, savedNumber is an NSString so everything's okay.) I'm going to assign what I get back to savedNumber, which is the instance variable I originally saved.

2. **Check to see whether the entry exists:**

   ```
   if (savedNumber == nil)
   ```

   objectForKey: either returns the object associated with the specified key, or, if the NSUserDefaults can't find data for the key or doesn't find the key at all, it returns nil. That's precisely what's going to happen the first time I run the program — there's nothing stored because I didn't store anything yet.

3. **If there is no data there, create a new entry in the `NSUserDefaults` database.**

   I know that's easy for me to say. Let me show you how to actually do it.

   a. Create a new dictionary.

   ```
   savedNumber = @"650 555 1212";
   NSDictionary *savedNumberDict =
        [NSDictionary dictionaryWithObject:savedNumber
             forKey:kNumberLocationKey];
   ```

   To use `objectForKey` (to read preference data) or `setObject` (to update preference data), you have to create an entry for the item you want to read or update in `standardUserDefaults`. (For `NSUserDefaults` to know anything about my preferences, you have to *tell* it about them first, right?) To let `NSUserDefaults` know what's going on, you have to create a dictionary listing of all the key-value pairs you plan on using here — all *one* of them, because all you need is the `savedNumberforKey`/`kNumberLocationKey` value pair.

   `dictionaryWithObject:forKey:` creates and returns a dictionary containing the key and value you give it. You pass it `savedNumber`, which you initialized with `650 555 1212` (the value), and `kNumberLocationKey` (the key). Notice I had you use the same value — `650 555 1212` — to initialize the preference that you used for the text field in Interface Builder back in Chapter 7, so as not to confuse the user.

   b. Register the defaults by using `registerDefaults:`.

   ```
   [[NSUserDefaults standardUserDefaults]
                    registerDefaults:savedNumberDict];
   ```

   `RegisterDefaults:` simply tells the `NSUserDefaults` object to add this key and this value to its database for this application. You have to do that only once, and then you can simply access it (using `objectForKey`) or update it (using `setObject`).

4. **If there's a saved number, I'm fine, and I can go along my merry way.**

Dictionaries manage pairs of keys and values. A key-value pair within a dictionary is called an *entry*. Each entry consists of one object that represents the key and a second object that is that key's value.

If you had more than one preference, you could have used `dictionaryWithObjectsAndKeys:`. That method creates and returns a dictionary containing entries constructed from the specified set of values and keys:

```
NSDictionary *dict =
    [NSDictionary dictionaryWithObjectsAndKeys:
        savedNumber, kNumberLocationKey, @"another value",
        @" another key", nil];
```

As you implement and experiment with this code, you need to be aware of the fact that you should delete the application from the simulator if you change anything of significance — the key, for example. The consequences of not doing so becomes obvious when things don't work like you expect them to. Deleting the application will delete any preferences for the app saved in `NSUserDefaults`.

Using data

There's only one thing left to do. The object that really cares about the number is the view, and it's the view controller's job to get it to the view. To put that saved data to use in an application's view, you have to link it up with a view controller — in this case, `ReturnMeToViewController`. If you look back at Figure 9-1, you can see that the best place to do that is `viewDid Load`, which is invoked right after the view has been loaded from the nib file. `viewDidLoad` is found in the `ReturnMeToViewController.m` file, so that's where you go to insert your code to do view initialization.

If you're ever lost in the file and need to find your next destination fast, use ⌘+F to open a Find dialog to find it, or use the drop-down menu I show you at the end of the Chapter 8.

Listing 9-7 shows the stuff you need to add to the `viewDidLoad` method in the `ReturnMeToViewController.m` file.

Listing 9-7:   Updating viewDidLoad

```
(void)viewDidLoad {

  textField.clearButtonMode =
                          UITextFieldViewModeWhileEditing;
  ReturnMeToAppDelegate *appDelegate =
              [[UIApplication sharedApplication] delegate];
  label.text = appDelegate.savedNumber;
  textField.text = appDelegate.savedNumber;

  [super viewDidLoad];
}
```

Here's what all the boldfaced stuff does:

1. **Gets the pointer to the application delegate object:**

   ```
   ReturnMeToAppDelegate *appDelegate =
               [[UIApplication sharedApplication] delegate];
   ```

   `sharedApplication` is a class method of `UIApplication` and returns the application delegate.

2. **Assigns the saved number to the label and text field:**

   ```
   label.text = appDelegate.savedNumber;
   textField.text = appDelegate.savedNumber;
   ```

appDelegate allows me to access the ReturnMeToAppDelegate's instance variable, savedNumber. As you know (because I mention it several times), one of the fundamental principles of object-oriented programming is *encapsulation* — tucking an object's instance variable behind a wall so you can't access it directly.

But earlier you did make it a property and you told the compiler to generate the necessary accessors in Listings 9-1 and 9-2, so accessors are available for you to use. Now, I could access savedNumber by using the getter:

```
[appDelegate savedNumber]
```

But as I mention in Chapter 7, I could also invoke an accessor method by using dot notation (which refugees from other object-oriented languages are sure to recognize):

```
appDelegate.savedNumber
```

Being one of those refugees myself, I use the dot notation.

For a method in ReturnMeToViewController to access the savedNumber instance variable, it needs to know where that variable is. That information is in the class declaration, in the ReturnMeToAppDelegate.h (header) file. So I need to use the #import ReturnMeToAppDelegate.h statement in the ReturnMeToViewController.m (implementation) file. This is shown in Listing 9-8.

Listing 9-8: Including the ReturnMeToAppDelegate Header File

```
#import "ReturnMeToViewController.h"
#import "ReturnMeToAppDelegate.h"
```

Finally, I need to clean up and de-allocate the memory. This is shown in Listing 9-9.

Listing 9-9: Releasing the New Variable

```
- (void)dealloc {
  [viewController release];
  [window release];
  [savedNumber release];

  [super dealloc];
}
```

Disabling Editing

I start this chapter by mentioning that I want to create a way that kept someone other than the main user from entering a new phone number after it had been initially entered. I can easily do that by changing an instance variable in the text field. In Figure 9-2, you can see that the UIView class, from which the UITextField class is derived, has a property instance variable userInteractionEnabled. (If Figure 9-2 looks different than your view, it's because I've hidden the table of contents by clicking the disclosure arrow.) If I set the userInteractionEnabled property of the UITextField to NO, the user can't enter any text; if the user taps the text field, nothing happens.

"But aren't you forgetting something?" you might ask. "Don't you have to allow the app's owner to be able change the number?"

I'm two steps ahead of you. To allow for just such a situation, I've decided to create a hidden "button." I thought it would be clever to make the label that displays the current number the hidden button, and to show you how flexible the framework is.

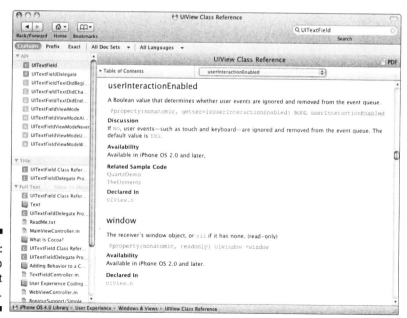

Figure 9-2: How to disable text entry.

A hidden button is only *one* way — not necessarily the *optimal* way from a user experience perspective — to provide the functionality you are after here. I'm starting with this technique to give you the basis for an understanding of how you can process touches on the iPhone. I also show you two additional ways in Chapter 11 (a traditional button and using a gesture) to accomplish the same functionality.

So how can I do that? To start with, labels are disabled for editing by default; they ignore touches. Your friend `userInteractionEnabled` is set to `NO` by default in a `UILabel`. All I need to do to enable touches in the label is change it to `YES`.

Again, the place to disable editing in the text field — and enable user interaction in the label — is in `viewDidLoad` (notice a pattern here?) in the `ReturnMeToViewController.m` file. The code for doing all this is shown in Listing 9-10, with the updates in bold.

Listing 9-10:   Updating viewDidLoad

```
- (void)viewDidLoad {

    textField.clearButtonMode =
                            UITextFieldViewModeWhileEditing;
    ReturnMeToAppDelegate *appDelegate =
            (ReturnMeToAppDelegate *)[[UIApplication
            sharedApplication] delegate];
    label.text = appDelegate.savedNumber;
    textField.text = appDelegate.savedNumber;

    if (![appDelegate.savedNumber isEqualToString:
                            @"650 555 1212"]) {
        textField.userInteractionEnabled = NO;
        label.userInteractionEnabled = YES;
        }
    [super viewDidLoad];
}
```

Here the code checks to see whether the user has ever entered a phone number — by determining whether the number is equal to the default value that you set in Interface Builder.

What you're doing in the code is checking to see whether the `savedNumber` is *not* equal to the default number — which is why you see the negation operator `!` used here. If it's not equal, then the user has entered a number, and you need to disable the text field and enable the label.

Notice the message:

```
[appDelegate.savedNumber isEqualToString:@"650 555 1212"])
```

savedNumber is an NSString object, and one of NSString's handy methods is isEqualToString:. This will return YES if the text values of two string objects are equal. Now, I can see where you might be tempted to do the compare as savedNumber == :@"650 555 1212". Unfortunately, what you would be doing here is comparing the string pointers, not the actual text values of the objects.

Letting the User Use the Secret Button

I mention in the preceding section that I want to put a mechanism in place for the owner of our ReturnMeTo application that allows him or her to unlock a text entry field for editing. That mechanism is going to rely on the Old Secret Button trick, where the label field gets transformed into a trigger mechanism for changing the text-entry field from read-only to read/write. As you might expect, this is going to require some coding. The place to do this coding turns out to be the same method I used to process touches on the screen so the user could dismiss the keyboard — namely, the touchesBegan:: method.

touchesBegan:: is a message sent when one or more fingers touches down in a view. The touch of a finger (or lifting it from the screen) adds a touch event to the application's event queue, where it's *encapsulated* — placed into — a UIEvent object. There's a UITouch object for each finger touching the screen, which enables you to track individual touches. As the user manipulates the screen with his or her fingers, the system reports the changes for each finger in the corresponding UITouch object.

Listing 9-11 shows how to determine whether the user has touched the label; if so, the text field is enabled so the user can enter a new number. (Don't forget the new closing brace after the existing if block.)

Listing 9-11:   Is the Touch in the Label?

```
- (void)touchesBegan:(NSSet *)touches withEvent:
                                        (UIEvent *) event {

   if (!textField.userInteractionEnabled) {
     UITouch *touch = [touches anyObject];
     if (CGRectContainsPoint([label frame],
                       [touch locationInView:self.view])) {
       textField. userInteractionEnabled = YES;
       label.text = @"You found it, touch below";
       textField.placeholder =
                       @"You may now enter the number";
   }
  }
```

(continued)

Listing 9-11 *(continued)*

```
else {

  if (textField.editing) {
    [textField resignFirstResponder];
    [self updateCallNumber];
    textField.userInteractionEnabled = NO;
  }
}
[super touchesBegan:touches withEvent:event];
}
```

Here's how the code builds its magic button, step by step:

1. **If the text field is not enabled, get the touch object.**

```
if (!textField. userInteractionEnabled) {
         UITouch *touch = [touches anyObject];
```

   Touches are passed in an `NSSet` object — an "unordered collection of distinct elements," for those of you not up on the intricacies of `NSSet` objects. To access an object in the `NSSet`, you use the `anyObject` method. This returns one of the objects in the set, or `nil` if the receiver contains no objects. Bear in mind that the object returned is chosen by some magic formula developed in secret by a cabal of Apple developers — the only thing *I* know about it is that the selection is not guaranteed to be random.

2. **Check to see whether the touch was in the label.**

```
if (CGRectContainsPoint([label frame],
          [touch locationInView:self.view])) {
```

   `CGRectContainsPoint` is a function that returns `YES` when a rectangle contains a specified point. You use the label `frame` here, which is a rectangle in the coordinates of the view — that thing on the screen that you, I, and the user see.

   Well, if I know where the view label is in the view's coordinate system, then I'd better know where the fingers are in the view's coordinate system as well. The problem is that if the user touches in a label, the OS could report back the touch's location in terms of the view or in terms of the label. Fortunately, I can specify the terms — the coordinate system, to be precise — I want.

```
[touch locationInView:self.view]
```

   This method returns *the current location of the touch* in the coordinate system of a given view — in this case, `self.view` specifies that you want the location of the touch in the content view's coordinate system. That means I'm comparing apples to apples. (Do you believe I actually said that?)

3. **If the touch is in the label frame, display a clever message, enable the text field, and set the placeholder to give the user more guidance.**

The placeholder is what the user sees when he or she touches the text field to start editing.

```
textField. userInteractionEnabled = YES;
label.text = @"You found it, touch below";
textField.placeholder =
                    @"You may now enter the number";
```

4. **Finally, you want to disable any further editing as you did in the earlier section, "Disabling Editing."**

```
textField.userInteractionEnabled = NO;
```

You also want to disable further editing after the user touches Return on the keyboard. To do that, add the line of code in bold in Listing 9-12 to `textFieldShouldReturn:`.

Listing 9-12:   Disabling Editing after the User Touches Return

```
-(BOOL)textFieldShouldReturn:(UITextField *)theTextField {

    [textField resignFirstResponder];
     [self updateCallNumber];
    textField.userInteractionEnabled = NO;
     return YES;
}
```

What You Have Now — At Long Last

So it looks like you have all the pieces in place for your ReturnMeTo application. You can now enter and save numbers, and keep someone from changing the number if he or she doesn't know how.

Appearances can be deceiving, though.

Reality check: Some how-to books on software development should really be housed in the Fiction section of your local bookstore — because all their examples work flawlessly. In the real world — the nonfictional world — everything does not always go as planned; occasionally your software program blows up on you. That's why an essential part of software development is the *debugging* phase — teasing as many flaws out of your app as possible so you can squash 'em. In the next chapter, I show you how to work through the debugging phase of your project and introduce you to the SDK's very own debugging tool, something that's sure to make your software-development life a lot easier.

For a nice retrospective of the work you've done so far for the ReturnMeTo application, move on over to my Web site (www.nealgoldstein.com) and check out Listings 9-13 through 9-16. There, in all their glory, you can find your app's completed code listings (to this point).

Chapter 10

Using the Debugger

*F*ace it: When you're developing an application, sometimes things don't work out quite the way you planned — especially when you knock over a can of Jolt Cola on the keyboard and fry it out of existence.

"Stuff happens," in the immortal words of a famous ex-U.S. Secretary of Defense. When it comes to developing your own programs, that "stuff" comes in three categories:

✔ **Syntax errors:** Compilers — the Objective-C compiler in Xcode is a case in point — expect you to use a certain set of instructions in your code; those instructions make up the language it understands. When you type `If` instead of `if`, or the subtler `[view release}` instead of `[view release]`, the compiler suddenly has no idea what you're talking about and generates a syntax error.

Syntax errors are the most obvious of errors out there, simply because your program won't compile (be able to run) until all of these are fixed. Generally, syntax errors spring from typographical errors like those mentioned here. (And yes, the errors can be pretty penny-ante stuff — an *I* for an *i*, for goodness sake — but it doesn't take much to stump a compiler.)

In Figure 10-1, you can see an example of a syntax error. This one was kindly pointed out to me by Xcode's friendly little Debugger feature (more on him later). I'd forgotten to put a semicolon (`;`) after `UITextFieldViewModeWhileEditing` in this line. It looked like this:

```
textField.clearButtonMode =
                    UITextFieldViewModeWhileEditing
```

As a result, I got a number of errors because the compiler couldn't quite figure out what I was doing. To get the error explanation that you see in Figure 10-1, I clicked the red exclamation point icon in the gutter to the left of the statement in Figure 10-1.

It's generally better to ignore the subsequent errors after the first syntax error because they may be the result of that first error. In this case, because of the first error, that line and the next one were treated as a single instruction, and the line that declared appDelegate wasn't processed by the compiler as a separate instruction.

When the compiler wants to inform you about an issue during a compile, you'll see an indication in the notification section (in the bottom-right corner) of the Xcode window. In the case of a failed compile, you see a tiny hammer icon and Failed, as you can see in Figure 10-2. If you click anything in the notification section, it takes you to the error in the Build Results window, as shown in Figure 10-2.

Although in this chapter I show you errors in the Xcode Text Editor window, I generally do my debugging in the Build Results window.

✔ **Runtime errors:** *Runtime errors* cause your program to stop executing — it "crashes," in other words, as in "crash and burn to much wailing and gnashing of teeth." Something might have come up in the data that you hadn't expected (a division-by-zero error, for example), or the result of a method dealt a nasty surprise to your logic, or you sent a message to an object that doesn't have that message implemented. Sometimes you even get some build warnings for these errors; often the application launches and then shuts down, simply stops working, or "hangs" (stops and does nothing).

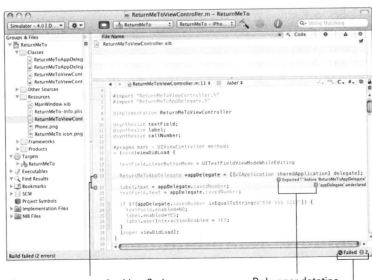

Figure 10-1:
A syntax error. Oops.

Syntax errors recognized by xCode

Debugger datatips

Notifications

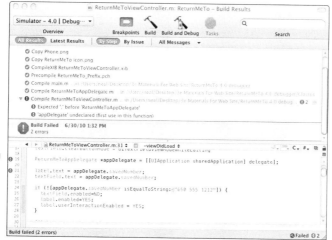

Figure 10-2:
The Build
Results
window.

✔ **Logic errors:** Your literal-minded application does exactly what you tell it to, but sometimes you unintentionally tell it the wrong thing, and it coughs up a *logic error*. In Figure 10-3, everything looks fine — not an error sign in sight — except when I try to enter a number into the text field, I discover that the field is disabled. Ironically, if I fool around and touch the label, I find that *then* I can enter text — which is the opposite of what I want.

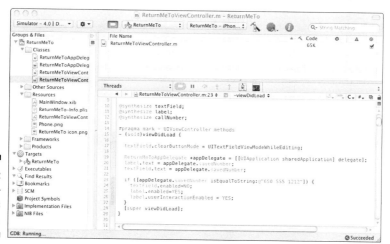

Figure 10-3:
Oh, great —
it works
backwards.

Look a bit more closely at this chunk of code, though:

```
if ([appDelegate.savedNumber
                isEqualToString:@"650 555 1212"]) {
    textField.enabled = NO;
    label.enabled = YES;
    label.userInteractionEnabled = YES;
}
```

See how I tell it to disable the text field if the saved number is `"650 555 1212"` but not to disable if the saved number is *not* equal to that value? I had mistakenly left out the `!` (now in bold) before the compare to the `appDelegate`'s `savedNumber`. Here's how it should have looked (note how it begins):

```
![appDelegate.savedNumber
            isEqualToString:@"650 555 1212"]
```

Not being able to guess what it was that I *really* wanted, it did what I *told* it to; the program worked, but not the way I intended it to.

Syntax errors, runtime errors, and logic errors can all be pains in the behind, but there's no need to think of them as insurmountable roadblocks. You're still on your way to a cool iPhone app. In this chapter, I show you how to use the Debugger to remove at least some of these obstacles. The Debugger works best for Case 2 (the runtime errors), but as I point out later on, it can also help you track down the logic errors in Case 3.

Using the Debugger

In Figure 10-4, I deliberately created a situation that will give me a runtime error: I'm going to divide by zero — a mathematical no-no that any able-bodied fifth-grader would berate me for — which will enable me to show you how to approach runtime errors in general.

Here's the drill: After introducing my boneheaded error while writing my code in Xcode, I clicked the Build and Run button in the toolbar.

The application started up and then immediately stopped, and as I explain in Chapter 4, the compiler was kind enough to tell me what I was doing wrong.

I started by clicking the warning icon (in this case a yellow exclamation point) in the gutter. You can see in the top of Figure 10-4 that I got a warning message `Division by zero` followed by a 2, which means another warning is there. In the bottom half of Figure 10-4, I then clicked the 2, and it revealed a second warning, `Unused variable I`. Pay attention to those numbers because sometimes the subsequent message can help you more precisely understand what the compiler is complaining about.

If you want to see these warnings in the Build Results window (as I do), all you need to do is click anything in the notification section.

Before you scoff and say that I should have caught such a basic error, okay, you're right to scoff — but I want to point out some things:

✔ In the middle of development, I may have been pelted with compiler warnings that I didn't really need to take care of because they had no impact on the execution of the program. As a result, I might not have noticed *one more compiler warning* that actually *did* have an impact — a big one.

✔ If my app was a bit more complicated, I could conceivably end up dividing by zero without realizing it. (Remember, stuff happens.)

✔ If you happen to save this code and then build it again, you won't see the compiler warning in the status bar because there were no changes to that file — so it wouldn't be recompiled. (This is definitely something to remember.) This is one reason to set the Xcode Building preference Build Results Window: Open during builds: to Always — it will continually remind you about those warnings.

How can the Debugger help me determine the source of a runtime error like this one? The next section gives you the details.

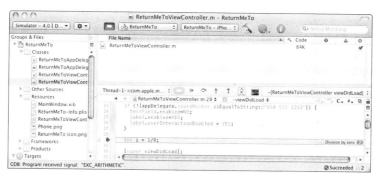

Figure 10-4:
Results of division by zero.

Debugging your project

As you may have noticed, when I do get an execution error, the Debugger strip becomes visible.

If you're running with Breakpoints on, as I suggest in Chapter 4, you're all set. If you aren't (the Build and Run button is green), click the Breakpoints button on the Project window toolbar. The Build and Run button changes to orange (if this book were in color, you'd see that in Figure 10-5). When everything is set up. I want to point your attention to a few things you may have noticed in Figure 10-4 — or if your program ever had an execution error.

1. You get a message in the Debugger Console as well as in the status bar of the Project window, as shown in Figure 10-5.

   ```
   Program received signal:  "EXC_ARITHMETIC".
   ```

2. The Debugger strip is visible in the Project window, just above the Editor view, as you can see in Figure 10-5. There are also a number of buttons for your pushing pleasure. I get to each of them shortly.

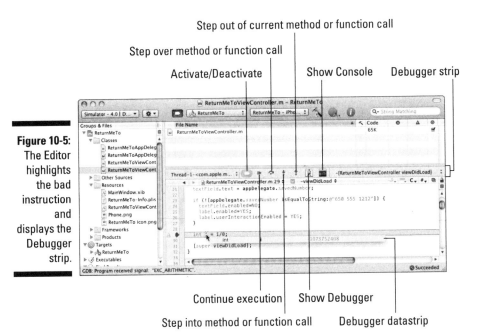

Figure 10-5:
The Editor highlights the bad instruction and displays the Debugger strip.

3. In Xcode's Editor view in Figure 10-5, notice the red arrow that shows you the instruction that caused the program to crash. That's the Debugger pointing out the problem to you.

4. There's even more information though. As you can see in Figure 10-5, I have my mouse pointer positioned above the i on the offending line. You can see the Debugger datatip displaying the value of i. What's even more interesting is that if I position my mouse pointer above appDelegate, shown in Figure 10-6, and then move it over the disclosure triangle, I get a Debugger datatip that shows me the appDelegate's instance variables — including savedNumber, which is 650 555 1212, which is what you would expect.

5. If you select the up and down arrows next to [ReturnMeToViewController viewDidLoad] in the Debugger strip in Figure 10-6, you get to see the stack — a trace of the objects and methods that got you to where you are now — as shown in Figure 10-7.

   For example, a quick look at Figure 10-7 shows that main called UIApplicationMain — which sent the UIApplication_run message, and so on, which eventually ended up in ReturnMeToAppDelegate: application:didFinishLaunchingWithOptions: and then finally to ReturnMeToViewController viewDidLoad. And that's where my little problem reared its ugly head.

   Okay, the stack isn't really all that useful in this particular context of dealing with my boneheaded attempt to divide by 0. But it *can* be very useful in other contexts. In a more complex application, the stack can help you understand the path that you took to get to where you are. Seeing how one object sent a message to another object — which sent a message to a third object — can be really helpful, especially if you didn't expect the program flow to work that way.

Figure 10-6:
A Debugger
datatip.

Getting a look at the stack can also be useful if you're trying to understand how the framework does its job, and in what order messages are sent. As you see later in this chapter, using something called a *breakpoint* can stop the execution of my program at any point and trace the messages sent up to that point. So don't despair; you have options.

There's even more information available, though — it comes to you in the Debugger window. The next section gives all the details, so read on.

Figure 10-7:
Looking at the stack in the Editor view.

Using the Debugger window

Even though the Debugger is officially running, you have to open the Debugger window explicitly the first time you choose Run⇨Debugger. To do that, you can click Show Debugger in the Debugger strip in the Project window, or choose Run⇨Debugger from Xcode's main menu (or press Shift+⌘+Y).

Figure 10-8 shows the Debugger window in all its glory. It has everything that is in the Editor window, but you can see your stack and the variables in scope at a glance. It also has some extra functionality I show you in the upcoming section, "Using Breakpoints."

Note that, in the upper-left pane, you can see the same stack you can see in the Editor.

Okay, your window may not look exactly like mine. That's because Xcode gives you lots of different ways to customize the look of the Debugger window. You could, for example, have chosen Run⇨Debugger Display from the main menu and tweaked the way you want your Debugger window to look. The option I chose was Source Only — so that only the source code appears in the bottom pane. You could, of course, have selected the Source and Disassembly option if you had a hankering for checking both the source code *and* the assembly language (if you really care about assembly language); in that case, the bottom pane would divide down the center into two panes, with the source code on the left and the assembly code on the right.

Figure 10-8:
The
Debugger
window.

It's not that I actually expect you to use the Source and Disassembly option — at this point, I don't. But sometimes, as you explore interfaces, things end up not looking the way they used to. In my experience, this usually occurs because a different display option has been chosen — either by accident or on purpose.

Examine the top-right pane in the Debugger window. There you see a display of the program object's variables. I clicked the disclosure triangles next to self and appDelegate, and you can see the appDelegate's instance variables that you saw in the datatip in the Editor window; for example, savedNumber, which is (still) 650 555 1212, which is what you would expect. You can also see as well the instance variables for the ReturnMeToViewController under self, which is under Arguments.

This is useful for a couple of reasons:

✔ **Checking variables:** If the view isn't displaying the correct number, I can look in the Variable pane to see what the value of the variable actually is. If the value is correct here, I can conclude that either it gets changed by mistake later, or I'm displaying something other than what I intended to display.

✔ **Checking messages sent:** Some logic errors you may encounter are the result of what some people call a "feature" and others call a "design flaw" in Objective-C: For reasons which are not particularly important here, Objective-C, unlike some other languages, allows you to send a message to a `nil` object *without* generating a runtime error. If you do that, you should expect to subsequently see some sort of logic error because a message to a `nil` object simply does nothing.

So, when things don't happen the way you expect, you might have a real logic error in your code. But there's one other possibility: Maybe an object reference has not been set, and you're sending the message into the ether.

Now, how can I use the variable pane to help me with that? Simple. If you look at an object reference instance variable and its value is 0x0, any messages to that object are simply ignored. So when I get a logic error, the first thing I'm going to check is whether any of the object references I am using have 0x0 as their values, informing me that the reference was never initialized.

✔ **Checking for initialization:** Notice the value of `i` in Figure 10-8. That long, seemingly random number is the way it is because it hasn't been initialized yet. The instruction that was going to initialize it was the one that generated a runtime error — because I tried to initialize `i` with the result of a divide-by-zero operation.

As you can see, the debugger can be really useful when your program isn't doing what you expect. For the blatant errors, the debugger can show you exactly what is going on when the error occurred. It provides you with a trail of how you got to where you are, highlights the problem instruction, and shows you your application's variables and their values at that point. Had the cause of the error in this case been more subtle, looking at the value of the variable would have given me a good hint about what was going on.

What's just as valuable is how the debugger can help you with logic errors. Sending a message to `nil` is not uncommon, especially when you're making changes to the user interface and forget to set up an outlet, for example. In such situations, the ability to look at the object references can really help. And what can really help you with that is the ability to set breakpoints, which is the subject of the next section.

Using Breakpoints

Xcode's Debugger feature is a great tool for tracking down runtime errors, as earlier sections in this chapter make clear. I want to highlight another useful feature of the Debugger: its capability of setting breakpoints. If you're stymied by a logic error, setting breakpoints is a great way to break that logjam.

A *breakpoint* is an instruction to the Debugger to stop execution at that instruction and wait for further instructions (no pun intended). By setting breakpoints at various methods in my program, I can step through its execution, at the instruction level, to see exactly what it's doing. I can also examine the variables the program is setting and using, which will allow me to determine if that's where the problem lies.

In Figure 10-9, I've set a breakpoint simply by clicking in the far-left column of the Editor window. (I also deleted the statement that did the division by zero.) When I build and run the program again, as you can see in the Debugger Editor window in Figure 10-10, the program stopped executing right at the breakpoint I set. (You would've also seen that same thing in the Editor window.)

In Figure 10-9, I've also clicked the triangle next to the appDelegate. That shows the appDelegate's variables, one of which is the viewController. If you notice the value under appDelegate for the viewController, it's 0x8010020. If you notice the value at the very top for self, you see it's also 0x8010020. This makes sense because the viewController variable is pointing to the viewController, which is in the object we are now in, and which is the value for self.

Figure 10-9:
Setting a
breakpoint.

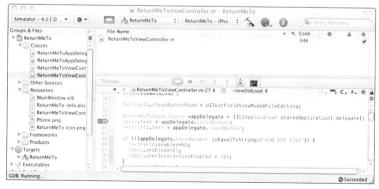

I've also clicked the triangle next to `self`, which shows me the view controller's variables. If I had clicked `viewController` under `appDelegate`, I would have obviously seen the same thing, or I could have stayed in the Editor and used the datatip.

Looking under the `appDelegate` variables, you can see the `callNumber` and its value, which is `nil`. If I'd been trying to do something by using `callNumber` at this point, that click would have shown me that I was trying to do something with a variable that hadn't been initialized with the value I needed. In other words, this would not have been a good time in the program flow to display the `callNumber` variable.

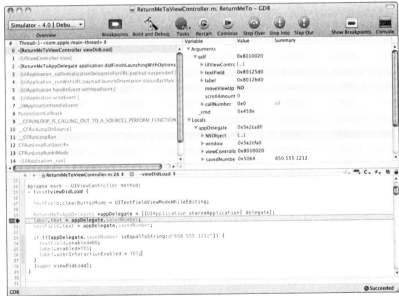

Figure 10-10:
What the Debugger window shows at the breakpoint.

Suppose I want to see precisely when that variable was set. I could execute the program instruction by instruction, simply by clicking the Step Into button on the Debugger toolbar. I would have executed `textField.text =appDelegate.savedNumber;` (after a brief stop at `@synthesize savedNumber` — I could have clicked Step Over and skipped that brief visit) and then gone on to the next instruction, as you can see in Figure 10-11. I'd keep on clicking that Step Into button at every instruction until I got to where I wanted to be (which, by the way, can be a long and winding road).

The Debugger window gives you a number of other options for making your way through your program in addition to Step Into. For example, you could use one of the following:

✔ **Step Over** gives you the opportunity to simply go to the next instruction without going through all the instructions in a method call if there is one.

✔ **Step Out** takes you out of the current method.

✔ **Continue** tells the program to keep on with its execution.

✔ **Restart** restarts the program. (You were hoping maybe if you tried it again it would work?)

To get rid of the breakpoint, simply drag it off to the side. You can also right-click the breakpoint and choose Remove Breakpoint from the pop-up menu that appears.

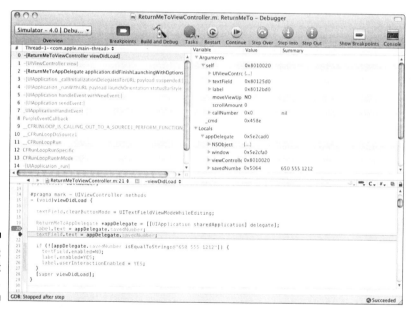

Figure 10-11:
The next
step.

Using the Static Analyzer

Xcode has a new Build and Analyze feature (the Static Analyzer) that analyzes your code.

The results show up like warnings and errors, with explanations of where and what the issue is. You can also see the flow of control of the (potential) problem. I say potential because the Static Analyzer can give you false positives.

In Figure 10-12, I deliberately created a memory leak. As you can see, I allocated a new `ReturnMeToViewController` and then did nothing with it.

```
ReturnMeToViewController* anObject =
                    [ReturnMeToViewController alloc];
```

I then chose Build and Analyze from the Build menu (Build⇨Build and Analyze). In Figure 10-13, you can see the results in the Project window. I get a warning (ignore the unused variable warning on line 19). I also get an analyzer result — the little blue icon in the gutter to the left of the line 21 number — that displayed the following when I clicked it:

```
Potential leak of an object allocated on line 19
```

Figure 10-12: A deliberate memory leak.

Figure 10-13: Running the Static Analyzer.

If I click the little blue icon the proceeds the "Potential leak" message, I get a "trace" of what happened in Figure 10-14.

Figure 10-14:
The
expanded
Static
Analyzer
warning.

First I get the warning:

```
Method returns an Objective-C object with a +1 retain
count (owning reference)
```

Then in the next line, it tells me

```
Object allocated on line 19 is no longer referenced after
this point after this point and has a retain count of +1
(object leaked)
```

Notice that the results refer to line numbers. That's why I made a point of explaining about how to turn on line numbers in Xcode in Chapter 4.

As I mention before, memory management is a big deal on the iPhone.

Before you attempt to get your application into the App Store, or even run it on anyone's iPhone, you need to make sure that it's behaving properly. By that I mean not only delivering the promised functionality, but also avoiding the unintentional misuse of iPhone resources. Keep in mind that the iPhone, even though it's a cool device, is nevertheless somewhat resource constrained when it comes to memory usage and battery life. Such restraints can have a direct effect on what you can (and can't) do in your application.

Although the Static Analyzer can help you detect memory leaks, the real champ at doing that is Xcode's Instruments application, which also lets you know how your application uses iPhone resources such as the CPU, memory, network, and so on.

The Instruments application allows you to observe the performance of your application while running it on the IPhone, and to a lesser extent, while running it on the Simulator. Here *instrument* means a specialized feature of the Instruments application that zeroes in on a particular aspect of your app's performance (such as memory usage, system load, disk usage, and the like) and measures it. What's really neat, however, is the fact that you can look at these different aspects simultaneously along a timeline — and then store data from multiple runs, so you get a picture of how your application's performance changes when you tune it.

Although the Instruments application is a very powerful piece of software, it has so many features that an in-depth discussion is beyond the scope of this book. If you're interested though, you can find an explanation in the *iPhone Application Development All-in-One For Dummies* by, yes, me again (Wiley). I'll leave that for you to explore on your own.

One More Step

Obviously, the Debugger is a very valuable tool — even more so because it's so easy to use. You can use the Debugger to figure out how your code and the framework are interacting at breakpoint, examine the stack, and figure out where the bug is hiding.

So where am I?

At this point, you can enter and save the number — and after the first time you do that, you can enter a new number only if you touch the number displayed in the label, which enables editing. (The code for doing that is in Chapter 9, Listing 9-11.)

Originally I was planning to stop here in the development of the ReturnMeTo application. But as I showed it on my iPhone to prospective users (okay, my friends with iPhones), they had a couple of suggestions that made me realize (after first arguing that it was okay the way it was) that I wasn't done:

- ✔ One suggestion was that you should be able to touch the phone number and have it dial automatically for you.

- ✔ The second was that touching the label display was kind of hokey, and a better way to do what I wanted to do was to change the iPhone image into a button.

I show you how to do both in the next chapter.

Chapter 11

Buttoning It Down and Calling Home

*D*evelopers tend to be optimistic when they put what they hope will be the finishing touches to their latest application. You've expended a certain amount of time and effort making your dream app a reality — okay, maybe not blood, sweat, and tears, but a lot of work nevertheless — and you tend to think that your efforts will be greeted with great praise.

The reality, more often than not, is that when you take your spanking-new application around to a few friends and colleagues it isn't always instantly welcomed as the newest, most advanced thing since sliced bread! Sure, it gets some praise, but there are some criticisms as well — "helpful suggestions" is how your friends put it.

To take a concrete example, when I started showing around the ReturnMeTo application the way it worked at the end of Chapter 9, I got the good-news-bad-news routine.

The good news was that everyone liked it. The bad news was, they almost all made the same two comments:

  ✔ Disabling editing was a good idea, but they also all thought that having to touch the label to enable it was pretty hokey. More than one person suggested, "Make the iPhone image a button".

  ✔ The second comment was, "Why can't I just touch the phone number to make a call?" ("It *is* a phone, right?" was usually the capstone to that comment.)

So be prepared for comments from the peanut gallery about how "this and this" and "such and such" is what your app needs to make it better/faster/ stronger. And be confident enough in your own skills as a developer to actually take such suggestions seriously, dispassionately evaluate each one, and incorporate the better ones into your work. I thought both comments about the ReturnMeTo application made a lot of sense, so I put my coder's cap back on and retooled the ReturnMeTo application so that the iPhone image acted as the enabling trigger for updating the phone number (as requested) and the iPhone "phoned home" when somebody touched the displayed phone number (as requested). This chapter shows the steps I had to go through in order to incorporate these suggestions.

Adding a Button to Your iPhone Interface

If you made your way through enough of the coding in this book to get to this point (Chapter 11), more than likely you won't have trouble adding a measly button to the interface. If you just remember that buttons use the Target-Action design pattern I talk about in Chapter 2, you can get your head around the whole button concept pretty quickly.

Ready? Set? Go!

The Target-Action pattern

You use the Target-Action pattern (see Chapter 2) to let your application know that a user has done something. When he or she taps a button, for example, your application is supposed to respond — which usually happens in due course because the button invokes some method that you specified in your code. The method that gets invoked is (usually) in the view controller that manages the view in which the particular button resides. In the case of our ReturnMeTo application, that view controller is `ReturnMeToViewController`.

In Figure 11-1, I have added a `UIControl` (a button). As you can see, whenever the user taps that button, it sends the `buttonPressed:` message to the `ReturnMeToViewController`.

End of story — well, at least for what happens when the button is tapped. Here, I'm only replacing one control (the label) with another control (the button). It really comes down to a plumbing issue, and makes the point that when you use the Model-View-Controller design pattern, making a user-interface change is almost trivial — it just requires a little replumbing in the controller. And, oh yes, a little bit of replumbing in Interface Builder as well.

Working through your button code

If you add a button to your interface, you need to add a method to your code to handle those times when somebody decides to actually tap the button — the button *action,* in other words.

Okay, it's time to start. First, you need to add the action method to the interface (as shown in Listing 11-1). You do this in the ReturnMeToViewController.h file.

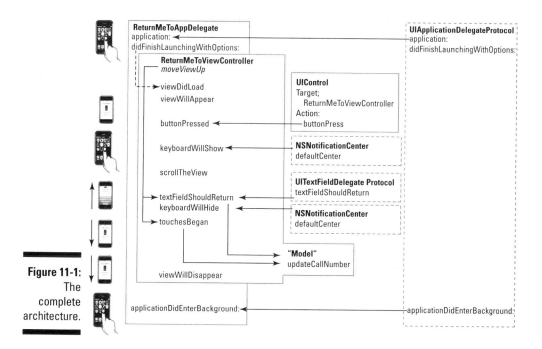

Figure 11-1:
The complete architecture.

Listing 11-1: Adding the Action

```
@interface ReturnMeToViewController : UIViewController
                                    <UITextFieldDelegate> {

  IBOutlet UITextField *textField;
  IBOutlet UILabel     *label;
  BOOL                  moveViewUp;
  CGFloat               scrollAmount;
  NSString             *callNumber;

}

@property (nonatomic, retain) UITextField *textField;
@property (nonatomic, retain) UILabel *label;
@property (nonatomic, retain) NSString *callNumber;

- (void)scrollTheView:(BOOL)movedUp;
- (void)updateCallNumber;
- (IBAction)buttonPressed:(id)sender;

@end
```

Here I've declared a new method — `buttonPressed` — with a new keyword right smack in front of it — `IBAction`.

`IBAction` is one of those cool little techniques, like `IBOutlet`, that does nothing in the code but provide a way to inform Interface Builder (hence, the `IB` in both of them) that this method can be used as an action for Target-Action connections. All `IBAction` "does" is act as a tag for Interface Builder — identifying this method (action) as one you can connect to an object (namely, the button) in a nib file. In this respect, this whole `IBAction` trick is similar to the `IBOutlet` mechanism I discuss in Chapter 7. In that case, however, you were tagging instance variables; in this case, methods. Same difference.

`IBAction` is actually defined as a `void`, so if you think about it, all you've done is declare a new method with a return type of `void`.

```
(IBAction)buttonPressed:(id)sender; =
```

```
(void)buttonPressed:(id)sender;
```

This simply means that you've declared a method that doesn't return anything to whoever invoked it.

The actual name you give the method can be anything you want, but it must have a return type of IBAction. Usually the action method takes one argument — typically defined as id, a pointer to the instance variables of an object — which is given the name sender. The control that triggers your action will use the sender argument to pass a reference to itself. So, for example, if your action method was invoked as the result of a button tap, the argument sender would contain a reference to the specific button that was tapped.

A word to the wise — having the sender argument contain a reference to the specific button that was tapped is a very handy mechanism, even if you're not going to take advantage of that in the ReturnMeTo application. With that reference in hand, you can access the variables of the control that was tapped.

Okay, you've declared the method; the next thing on your To-Do list is to actually add the buttonPressed method to the implementation file, Return MeToViewController.m. You can see the code that does this deed in Listing 11-2. I added a new #pragma here, and I added the method after that.

```
#pragma mark - Target Action methods
```

Just to keep things in the order I like, I added it right before

```
#pragma mark - "Model" methods

- (void)updateCallNumber {
```

I do that because I like to keep my "model" methods separated from overridden and delegate methods.

Listing 11-2:   Adding buttonPressed:

```
- (IBAction)buttonPressed:(id)sender {

    textField.userInteractionEnabled = YES;
    label.text = @"You found it, touch below";

    textField.placeholder =
                        @"You may now enter the number";
}
```

If this code looks familiar to you, you get extra points for paying attention. None of this code is new; I just moved it from the `touchesBegan::` method I put together back in Chapter 9, where I tied this code to the label. Remember, I said that user-interface changes were mostly going to be a matter of plumbing? Which reminds me — as long as you're going the button route instead of the label route, you'd best remove the code from the `touchesBegan::` method. To see what code needs to be excised, check out Listing 11-3, where I've commented out (and used `double strikethrough`) to show the code to be removed.

Listing 11-3:   Removing Code from touchesBegan::

```
- (void)touchesBegan::(NSSet *) touches
                         withEvent:(UIEvent *) event {

// if (!textField.userInteractionEnabled) {
//    UITouch *touch = [touches anyObject];
//    if (CGRectContainsPoint([label frame], [touch
//              locationInView:self.view])) {
//      textField.userInteractionEnabled = YES;
//      label.text = @"You found it, touch below";
//      textField.placeholder = @"You may now enter the
//              number";
//    }
//  }
//  else {

  if( textField.editing) {
    [textField resignFirstResponder];
    [self updateCallNumber];
    if (moveViewUp) [self scrollTheView:NO];
    textField.userInteractionEnabled = NO;
  }
//}

    [super touchesBegan:touches withEvent:event];
}
```

One last bit of cleanup and you're through with this chore. Because you no longer want the label to act as a trigger for enabling editing in the ReturnMeTo application, you have to remove the code for enabling the label that you added to `viewDidLoad` back in Chapter 9. Listing 11-4 shows what needs to go. (Again, look for the lines that have been commented out and have double strikethrough.)

Listing 11-4:   Removing the Label-Enabling Code from viewDidLoad

```
- (void)viewDidLoad {

  textField.clearButtonMode =
                          UITextFieldViewModeWhileEditing;

  ReturnMeToAppDelegate *appDelegate =
            [[UIApplication sharedApplication] delegate];
  label.text = appDelegate.savedNumber;
  textField.text =appDelegate.savedNumber;
  if (![appDelegate.savedNumber isEqualToString:@"650 555
        1212"]) {
    textField.userInteractionEnabled = NO;
// label.userInteractionEnabled = YES;
  }
  [super viewDidLoad];
}
```

Adding a button to your interface really takes no more code than that. Getting the button to actually *do* something for you takes a bit more work, but (fortunately) Interface Builder makes that task pretty much a snap. But before you go there, be sure to save your work. You know the drill: ⌘+S.

Connecting the Button in Interface Builder

So far, you've implemented the method you want to have invoked when the user taps the iPhone image button. Now, you need to do two things to make that work:

1. Create the button.

2. Give the button a method to invoke and tell it the object that method is in.

Doing this is really easy thanks to your (now) old buddy Interface Builder. If you think about it, you're doing the same thing you did when you connected the outlets in Chapter 7. Now you connect the action.

Launch Xcode and open the ReturnMeTo project:

1. **In the Groups & Files list on the left, double-click the ReturnMeToViewController.xib file.**

   You'll find the file in the Resources folder of the ReturnMeTo project's main folder.

   Double-clicking the file launches Interface Builder, which should display the three windows you see in Figure 11-2.

   Note: If the Library window isn't open for some reason, open it by choosing Tools➪Library from the main menu. And (weirder still) if the View window isn't visible, open it by double-clicking the View icon in the `ReturnMeToViewController.xib` window. (Where there's a will, there's a way. . . .)

2. **Drag the Round Rect Button item from the Library window to the View window, placing the item right next to the iPhone image you added back in Chapter 5.**

   If you use the blue guide lines that Interface Builder provides, you can resize and position the button so it's the same size as your iPhone image, as I do in Figure 11-2.

3. **Delete the Image View your old iPhone image — by selecting it and pressing the Delete key and then drag the Round Rect Button item to where the Image View used to be.**

Figure 11-2:
Dragging a
button.

4. **Click to select your Round Rect Button in the View window and then choose Tools⇨Attribute Inspector to bring up the Attribute Inspector.**

Make sure that you have the button selected.

You're going to use the Attribute Inspector to add an image to the button.

5. **Choose Phone.png from the Image drop-down menu in the Attribute Inspector (see Figure 11-3) and then choose Custom from the Type drop-down menu (again, see Figure 11-3). Also, choose Layout⇨Size to Fit.**

Your old iPhone image makes a reappearance — this time as a button. And that's really it — that's all it takes to create a button using an image! Now to connect things up.

6. **Right-click the button to display a list of connections.**

7. **Drag from the little circle to the right of the Touch Up Inside item and drop it on the File's Owner icon in the `ReturnMeToViewController.xib` window, as shown in Figure 11-4.**

The Touch Up Inside connection is a good choice in this situation because Touch Up Inside is the event that is generated when the last place the user touched before lifting his or her finger was inside the button. This allows a user to change his or her mind about touching the button by moving his or her finger off the button before lifting it up.

Figure 11-3:
Adding an image to the button.

The blue guide line you see here when dragging is the same blue line you see when connecting Outlets, as you did in Chapter 7.

8. **With the cursor still over the File's Owner icon, let go of the mouse button and then choose buttonPressed from the pop-up menu that appears, as shown in Figure 11-5.**

Doing so makes your connection.

As you can see, Interface Builder found the IBAction tag you declared in the previous section and displays it for you as a choice. You can now rest easy.

If you save your work and then compile and run the application in the Simulator, you'll discover that the iPhone image is now a fully functioning button that works the same as touching the label had previously. And it's so much prettier.

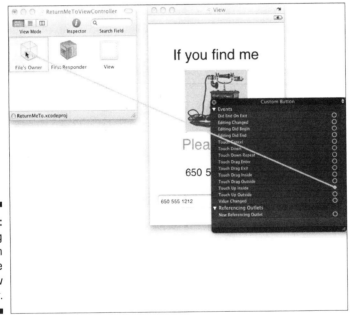

Figure 11-4:
Connecting the button with the view controller.

Figure 11-5:
Finishing the connection.

Phoning Home

Why push ten buttons when you can push just one? I had originally designed my ReturnMeTo application so that any Good Samaritan who found my lost iPhone could see the right contact phone number and give me a call so we could arrange a handover. When I showed this application to my friends, one in particular looked at me a bit incredulously when I told him he couldn't just press the displayed phone number to make the call. "Why not?" he asked. "Lots of applications do that."

I realized that he was right, and it's pretty easy to do that using a *Web view* (UIWebView), a view class that is part of the UIKit. It already has the ability to automatically recognize a phone number and initiate a call.

You use the UIWebView class not only to embed Web content in your application, but also to display anything coded with HTML — making it the number-one choice for displaying formatted text in your app. To do so, you simply create a UIWebView object. You can have the view take up the whole window (as I do in the iPhoneTravel411 application in Chapter 16), or you can also treat it like a control and make it a subview of your content view (which is functionally equivalent to how you used the label), as I do here. In either case, all you need to do is tell it to display some HTML content, and presto chango, you're done. That HTML content can be virtually (no pun intended) anywhere — on the Internet, stored locally, or in a string you create in your program. I show you how to deal with HTML content from the Internet and stored locally when I walk you through the iPhoneTravel411 application later on (in Chapters 13 through 18), but for now, I'm just interested in loading an HTML string — my beloved phone number.

The thing for you to remember is that, by default, a Web view recognizes a phone number and converts that number to a Phone Link. When the user taps a Phone Link, the Phone application launches and dials the number — which is precisely what you want.

Adding the Web view

Our ReturnMeTo application has managed so far with just a label view, simply because you only wanted the label to display a telephone number. Now that you want the number dialed when it's touched, you're going to replace the label view you're currently using with a Web view. To do that successfully, you need to tackle the following three tasks:

1. Identify and modify all the places that assign the telephone number to the label. Right now, to display the number in the label, all you have to do is assign it to label.text, and it gets displayed. It will now take a little more work to display it (but it's worth the effort) because you'll

have to replace each assignment with a message to the `UIWebView` to load the HTML string into `webView`.

2. Create the HTML string you want displayed.

3. Replace the `UILabel` with a `UIWebView` in Interface Builder and then connect everything.

Listing 11-5 shows the code you need. I started by making the necessary changes to the interface in the `ReturnMeToViewController.h` file. I replaced the `label` outlet with a `webView` outlet, and I also added an instance variable `htmlString` to compose the text I load into the Web view. The changes are in bold; I have commented out (and marked with `double strikethrough`) the deletions.

Listing 11-5: Updating the Interface

```
@interface ReturnMeToViewController : UIViewController
         <UITextFieldDelegate> {

         IBOutlet UITextField *textField;
//       IBOutlet UILabel      *label;
         IBOutlet UIWebView    *webView;
         BOOL         moveViewUp;
         CGFloat      scrollAmount;
         NSString     *callNumber;
         NSString     *htmlString;

}

@property (nonatomic, retain) UITextField *textField;
// @property (nonatomic, retain) UILabel *label;
@property (nonatomic, retain) UIWebView *webView;
@property (nonatomic, retain) NSString *callNumber;

- (void)scrollTheView:(BOOL)movedUp;
- (void)updateCallNumber;
- (IBAction)buttonPressed:(id)sender;
@end
```

You'll also have to add

```
@synthesize webView;
```

You should do that right after `@synthesize textField;` in the `ReturnMeToViewController.m` file. And because you have deleted the label, you're also going to have to delete

```
@synthesize label;
```

Dealing with this `@synthesize` business is shown in Listing 11-6.

Listing 11-6: Synthesizing the webView Instance Variable

```
#import "ReturnMeToViewController.h"
#import "ReturnMeToAppDelegate.h"

@implementation ReturnMeToViewController


@synthesize textField;
// @synthesize label;
@synthesize webView;
@synthesize callNumber;
```

Implementing the Web view

Implementing the Web view itself is fairly simple.

In the first version of the ReturnMeTo application, you simply assigned the `savedNumber` instance variable (the instance variable you added in Chapter 9 to hold the phone number you wanted to save) to the `label.text` field.

```
label.text = appDelegate.savedNumber;
```

As promised, moving to a Web view means you're going to have to do a little more work. In fact, you need to replace the `label.text` assignment with comparable code to create an HTML string and then have the Web view load and display it. The changes are in bold, and I've commented out (and marked with `double strikethrough`) the deletions.

There's one line in Listing 11-7

```
// label.userInteractionEnabled = YES;
```

that you previously deleted back in Listing 11-4. I left it in here so you wouldn't wonder where it disappeared to.

All of this involves existing methods in the `ReturnMeToViewController.m` file.

Listing 11-7: Changing viewDidLoad

```
- (void)viewDidLoad {


  textField.clearButtonMode =
                    UITextFieldViewModeWhileEditing;

  ReturnMeToAppDelegate *appDelegate =
```

(continued)

Listing 11-7 *(continued)*

```
            [[UIApplication sharedApplication] delegate];
// label.text = appDelegate.savedNumber;
   textField.text =appDelegate.savedNumber;
   htmlString = @"<div style=\"font-family:Helvetica,
               Arial, sans-serif; font-size:14pt;\"
               align=\"center\">";
   htmlString = [htmlString stringByAppendingString:
               appDelegate.savedNumber];
   htmlString = [htmlString stringByAppendingString:
               @"</span>"];

   [webView loadHTMLString:htmlString baseURL:nil];


   if (![appDelegate.savedNumber isEqualToString:@"650 555
         1212"]) {
     textField.userInteractionEnabled = NO;
// label.userInteractionEnabled = YES;
   }
   [super viewDidLoad];
}
```

All you're doing here is getting rid of the label assignment, the one that assigns the savedNumber "vanilla" string:

```
label.text = appDelegate.savedNumber;
```

And replacing it with code that creates an HTML-formatted string.

To start, you need to create a formatted HTML string, which surrounds the same appDelegate.savedNumber with the formatting information. Notice the appDelegate.savedNumber buried in there:

```
htmlString = @"<div style=\"font-family:Helvetica,
            Arial, sans-serif; font-size:14pt;\"
            align=\"center\">";
htmlString = [htmlString stringByAppendingString:
            appDelegate.savedNumber];
htmlString = [htmlString stringByAppendingString:
            @"</span>"];
```

To create the HTML string with the saved number in it, I use the stringBy AppendingString: method — which creates a string by adding one string after another. The only tricky thing here is that I had to use the backslash character before the embedded quotes in the HTML string.

I won't go into the HTML here. If you want to delve farther into HTML, feel free to check out *HTML, XHTML, & CSS For Dummies* by Ed Tittel and Jeff Noble (Wiley).

Then I simply tell the Web view to load the HTML string:

```
[webView loadHTMLString:htmlString baseURL:nil];
```

Here the `loadHTMLString::` method does exactly what I need it to do: It takes a string (with HTML formatting information embedded) and loads it into the Web view. Here's a closer look at the method's arguments:

✔ The first argument, `htmlString`, is the string I just formatted. Nothing surprising there.

✔ The second argument, `baseURL` is an `NSURL` object. An `NSURL` object, to no one's surprise, is an object that contains a URL. It can reference either a Web site or a local file. (It's not necessary here because you're not telling the Web view to access a URL.)

In effect, Listings 11-7, 11-8, and 11-9 all show you doing the same thing — adding the HTML string — in three different places.

Even though the HTML strings in the listings in this book span multiple lines, you can't do that in Xcode. You get errors such as

```
error: syntax error before'@' token
```

and

```
error: syntax error at 'OTHER' token
```

unless, of course, you use the escape character (\) as the last character on the line, right before the carriage return.

Listing 11-8:   Changing buttonPressed:

```
#pragma mark - Target Action methods

-(IBAction)buttonPressed:(id)sender{

  textField.userInteractionEnabled = YES;

  // label.text = @"You found it, touch below";

  htmlString =  @"<div style=\"font-family:Helvetica,
          Arial, sans-serif; font-size:14pt;\"
          align=\"center\">";
```

(continued)

Listing 11-8 *(continued)*

```
htmlString =  [ htmlString stringByAppendingString:
        @"You found it, touch below"];
htmlString =  [ htmlString stringByAppendingString: @"</
        span>"];
[webView loadHTMLString:htmlString baseURL:nil];


textField.placeholder = @"You may now enter the number";
}
```

Listing 11-9:   Changing updateCallNumber

```
- (void)updateCallNumber {

self.callNumber = textField.text;
// label.text = self.callNumber;
htmlString =  @"<div style=\"font-family:Helvetica,
        Arial, sans-serif; font-size:14pt;\"
        align=\"center\">";
htmlString =  [ htmlString stringByAppendingString:
        callNumber];
htmlString =  [ htmlString stringByAppendingString: @"</
        span>"];
[webView loadHTMLString:htmlString baseURL:nil];


ReturnMeToAppDelegate *appDelegate =
          [[UIApplication sharedApplication] delegate];
appDelegate.savedNumber = self.callNumber;
}
```

And don't forget that wonderful sentiment, "If you love something, set it free." In programming terms, that means you always have to release all the objects you own, as shown in Listing 11-10. (If this "release" stuff doesn't ring a bell, check out Chapters 6 and 7, where I cover the idea in greater detail.)

Listing 11-10:   Last but Not Least — dealloc

```
- (void)dealloc {

[textField release];
// [label release];
[webView release];
[callNumber release];
[htmlString release];

[super dealloc];
}
```

Adding and connecting the Web view in Interface Builder

As was the case when you replaced the Image view (your old iPhone image) with the Round Rect Button earlier in the chapter, you have to tell Interface Builder about our decision to replace the Label with a Web view. To get that ball rolling, do the following:

1. **Launch Xcode and open the ReturnMeTo project.**

2. **In the Groups & Files list on the left, double-click the `ReturnMeToViewController.xib` file.**

   Interface Builder launches, and you end up with the windows you see in Figure 11-6.

   Note: Again, if the Library window is not open for some reason, go ahead and open it by choosing Tools⇨Library from the main menu. And (weirder still) if the View window isn't visible, open it by double-clicking the View icon in the `ReturnMeToViewController.xib` window.

3. **Drag a Web view item from the Library window to the View window.**

   Use the blue lines to resize and position the Web view so it's the same size (and in the same place) as the label, as shown in Figure 11-6.

4. **Delete the (old) label by selecting it and pressing delete.**

5. **Right-click the File's Owner icon in the `ReturnMeToViewController.xib` window to display a list of connections.**

   Notice that this method of making a connection is different from the one I show you earlier in the chapter, when you had to deal with the Round Rect Button. I felt it would be a good idea to show you both ways of making the connection.

6. **Click webView in the list of connections — it's there under the Outlets heading — and drag the little circle there on the right over to the View window and onto the Web view itself, as shown in Figure 11-7.**

7. **Make sure that the Detects Phone Numbers check box is selected in the Attributes Inspector.**

Here you've replaced the label with a Web view and connected the new `webView` outlet to the Web view in the same way that the `label` outlet was previously connected to the label. (Not bad for a day's work.)

Figure 11-6:
Dragging a
Web view.

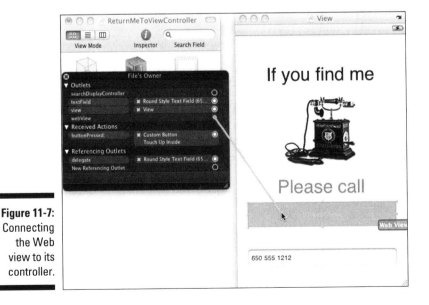

Figure 11-7:
Connecting
the Web
view to its
controller.

A Bug

In the spirit of making this a real-world exercise and not some piece of theory, I have to confess that when I was testing this application for the final time, three days before the final, final chapters were due, I found a bug. It doesn't get much more real world than that, does it?

When I replaced the label with the Web view, there was an unintended consequence (the polite way of saying I introduced a logic error in my code).

The way that I handled dismissing the keyboard and restoring the view when the user touched the screen (see Chapter 8) was by capturing the touches that were passed to the view controller in `touchesBegan::`.

That even worked if the user touched the label, because the label passed the touches on. But while the label doesn't care about touches, the Web view does. (It captures touches so it can detect when you touch a Web or Phone Link and load the URL or initiate the dialing of the number.) That meant that if the user happened to touch the screen in the Web view I just added — say, to indicate that he or she was done entering the number — `touchesBegan::` in the controller would never be invoked.

Ironically, the keyboard was dismissed because the Web view becomes the first responder when the user touches in it. (You can find out more about that in Chapter 8.) But because `touchesBegan::` was never invoked, the view was never restored to its pre-scrolled state.

Incidentally, I figured all this out by using a breakpoint in the handy-dandy Debugger — like I show you in Chapter 10. Talk about eating your own dog food.

Fixing it was simple. All I had to do was keep the Web view from handling touches during the period from when the user started using the keyboard until after the user dismissed it. I did that by setting the Web view's `user InteractionEnabled` property to `NO` as soon as the keyboard appeared. The place to do that was in `keyboardWillShow:` block object, which is invoked right before the keyboard appears (as you recall from Figure 11-1). All I had to do was add one line of code (in bold in Listing 11-11) to `key boardWillShow:`.

```
webView.userInteractionEnabled = NO;
```

Of course, after I disabled the Web view, I had to enable it again if I wanted the Web view to be able to initiate the call when the Phone Link was tapped. As you recall, `textFieldShouldReturn:` is invoked after the user has finished entering the phone number and taps Return, and `touchesBegan::` is invoked after the user has finished entering the phone number and taps the screen. So I added the following code (in bold in Listing 11-11) to both of those methods after the keyboard is dismissed.

```
webView.userInteractionEnabled = YES;
```

Piece of cake.

Listing 11-11:   Disabling and Re-Enabling the Web View

```
void (^keyBoardWillShow) (NSNotification *)=
                            (NSNotification * notif) {
    NSDictionary* info = [notif userInfo];
    NSValue* aValue =
        [info objectForKey:UIKeyboardFrameEndUserInfoKey];
    CGSize keyboardSize = [aValue CGRectValue].size;
    float bottomPoint = (textField.frame.origin.y +
          textField.frame.size.height + 10);
    scrollAmount = keyboardSize.height - (self.view.frame.
          size.height - bottomPoint);

    if (scrollAmount > 0)  {
      moveViewUp =YES;
      [self scrollTheView:YES];
    }
    else
      moveViewUp = NO;
    webView.userInteractionEnabled = NO;
  };


- (void)touchesBegan:(NSSet *)touches withEvent:
                            (UIEvent *) event {

// if (!textField.enabled) {
//    UITouch *touch = [touches anyObject];
//    if (CGRectContainsPoint([label frame], [touch
//        locationInView:self.view])) {
//      textField.userInteractionEnabled = YES;
//      label.text = @"You found it, touch below";
//      textField.placeholder = @"You may now enter the
//        number";
//    }
```

```
// +
// else {

  if( textField.editing) {
  [textField resignFirstResponder];
  [self updateCallNumber];
    if (moveViewUp) [self scrollTheView:NO];
    textField.userInteractionEnabled = NO;
    webView.userInteractionEnabled = YES;
  }
// +

  [super touchesBegan:touches withEvent:event];
}

-(BOOL)textFieldShouldReturn:(UITextField *)
                                        theTextField {

  [textField resignFirstResponder];
  webView.userInteractionEnabled=YES;
  if (moveViewUp) [self scrollTheView:NO];
  [self updateCallNumber];
  textField.userInteractionEnabled = NO;
  webView.userInteractionEnabled = YES;

  return YES;
}
```

Getting Even More In Touch

One of the great features of the iPhone is the Multi-Touch interface. You found out how to track touches in Chapter 8 when you allowed the user to dismiss the keyboard by touching anywhere on the screen. In Chapter 8, you also took that one step farther by tracking touches in a label. Now it's time to kick it up a notch by taking advantage of gestures on the iPhone.

In this section, I show you how to allow a user to use a pinch-out gesture to enable him or her to add a new phone number.

iPhone OS 4.0 provides you with a really easy way to do that. The UUIGestureRecognizer class and subclasses recognize gestures and even allow customizing their actions. The UIKit framework provides six gesture recognizers for the most common gestures, as shown in Table 11-1.

Table 11-1	Gesture Recognizers
Gesture	**Class**
Tapping (any number of taps)	`UITapGestureRecognizer`
Pinching in and out (for zooming a view)	`UIPinchGestureRecognizer`
Panning or dragging	`UIPanGestureRecognizer`
Swiping (in any direction)	`UISwipeGestureRecognizer`
Rotating (fingers moving in opposite directions)	`UIRotationGestureRecognizer`
Long press (also known as "touch and hold")	`UILongPressGestureRecognizer`

Before you decide to use a gesture recognizer, consider how you're going to use it. Your app should respond to gestures only in ways that users expect. In this case, even though a pinch out is usually used for zooming a view, one could also think about it as opening something up, like a text field. Okay, it's a stretch, but I like the idea anyway and even if you don't, you'll have discovered how easy it is to recognize gestures.

Start by creating a `UIPinchGestureRecognizer`. In Listing 11-12, add the code in bold to `viewDidLoad`.

Listing 11-12: Adding a Pinch Gesture Recognizer

```
- (void)viewDidLoad {

    textField.clearButtonMode =
                        UITextFieldViewModeWhileEditing;
    ReturnMeToAppDelegate *appDelegate =
                [[UIApplication sharedApplication] delegate];
    textField.text =appDelegate.savedNumber;
    htmlString = @"<div style=\"font-family:Helvetica,
            Arial, sans-serif; font-
            size:48pt;\"align=\"center\">";
    htmlString = [htmlString stringByAppendingString:
                appDelegate.savedNumber];
    htmlString = [htmlString stringByAppendingString:
                @"</span>"];
    [webView loadHTMLString:htmlString baseURL:nil];

    UIPinchGestureRecognizer *pinchGesture =
      [[UIPinchGestureRecognizer alloc] initWithTarget:self
                    action:@selector(handlePinchGesture:)];
    [self.view addGestureRecognizer:pinchGesture];
    [pinchGesture release];

    [super viewDidLoad];
}
```

Here you start by allocating the kind of gesture recognizer you want — in this case, a `UIPinchGestureRecognizer` — and initializing it with a target object (`self` or the `ReturnMeToViewController` object) and a method of your own choosing that's prepared to handle the gesture (`handlePinch Gesture:`). This method will be invoked whether it's a pinch in or out, and I show you in a second how to differentiate between the two.

```
UIPinchGestureRecognizer *pinchGesture =
    [[UIPinchGestureRecognizer alloc] initWithTarget:self
                    action:@selector(handlePinchGesture:)];
```

You add the gesture recognizer to the view

```
[self.view addGestureRecognizer:pinchGesture];
```

and then release it.

Gesture recognizers are attached to views, and although not part of the view hierarchy or responder, they do get first dibs on the touch and can do whatever they like with it. (I don't get into all of that here, but it's worth exploring on your own.)

A `UIPinchGestureRecognizer` knows exactly how to handle a pinch in or pinch out, and when one occurs it sends a `handlePinchGesture:` message to the `ReturnMeToViewController` object.

With your gesture recognizer in place, it's now your job to write the method you want to handle the gesture. Add the code in Listing 11-13 to `ReturnMeToViewController.m`.

Listing 11-13:   Handling the Pinch Gesture

```
- (IBAction)handlePinchGesture:
                          (UIGestureRecognizer *)sender {

    if (((UIPinchGestureRecognizer *)sender).scale >1.0)
      [self buttonPressed:self];
}
```

It really doesn't get much easier than this.

`UIPinchGestureRecognizer` has a `scale` property. It is the scale factor relative to the points of the two touches in screen coordinates. In other words, if the scale is greater than 1.0, the pinch is a pinch out, if it's less that 1.0 it's a pinch in.

In this case, you only want pinch out, so if the scale is greater than 1.0, you want to enable the text entry. Well, it just so happens that you've already written a method that knows exactly how to do just that: `buttonPressed:`.

So to save yourself some work — and the possible consequences of redundancy — you send yourself that message.

Almost done here. You also should adopt the `UIGestureRecognizerDelegate` protocol, which makes all this happen so easily.

To do so, add the code in bold to the `@interface` statement in `ReturnMeToViewController.h`. (Don't forget the comma before it.)

```
@interface ReturnMeToViewController : UIViewController
        <UITextFieldDelegate, UIGestureRecognizerDelegate>
```

You're Finally Done

With this last change, I'm putting the ReturnMeTo application to bed.

Chapter 12 — my All-Things-Having-to-Do-with-Provisioning-And-App-Store — will be the capstone to Part III. After that, you still have Part IV to deal with, where you get to look at some of the more advanced application-development techniques. There, I concentrate on explaining the ins and outs of designing my MobileTravel411 and iPhoneTravel411 applications. Drilling down a bit, I go through how to develop the iPhoneTravel411 app. Now, that's a more complex application — incorporating some techniques and tools you'll likely end up using in your own applications — stuff like saving files, using table views (to navigate, among other things), forming an alliance with the Settings application, figuring out internationalization, and accessing data on the Internet.

The Final Code

At this point in your career, what you should be interested in is not the listings you find in a book, but the Xcode project itself. For the most part, that's the way you'll be looking at sample code as well as code from other people you may be working with.

In that spirit, you can find the ReturnMeTo Xcode project on my Web site — www.nealgoldstein.com.

Chapter 12

Death, Taxes, and the iPhone Provisioning

*B*enjamin Franklin once said, "In this world nothing can be said to be certain, except death and taxes." I've discovered one other certainty in this earthly vale of tears: Everybody has the same hoops to jump through to get an application (1) onto an iPhone and then (2) into the App Store — and nobody much likes them, but there they are.

So you're working on your application, running it in the Simulator, as happy as a virtual clam, and all of a sudden you think you're ready to get it into the App Store. The first hurdle is getting the app to run on the phone.

For most developers, getting their applications to run on the iPhone during development can be one of the most frustrating things about developing software for the iPhone. The sticking point has to do with a rather technical concept called *code signing,* a rather complicated process designed to ensure the integrity of the code and positively identify the code's originator. Apple requires all iPhone applications to be digitally signed with a signing certificate — one issued by Apple to a registered iPhone developer — before the application can be run on a development system and before they're submitted to Apple for distribution. This signature authenticates the identity of the developer of the application and ensures that there have been no changes to the application after it was signed. As to why this is a big deal, here's the short and sweet answer: Code signing is your way of guaranteeing that no bad guys have done anything to the code that can harm the innocent user.

Now, as I said, nobody really likes the process, but it's doable. In this chapter, I start by giving you an overview of how it all works by jumping right to that point where you're getting your application ready to be uploaded to the App Store and then distributed to end users. I realize I'm starting at the end of the process, which for all practical purposes begins with getting your application to run on a device during development. I do the overview in this order because the hoops you have to jump through to get an application to run on a single iPhone during development are a direct consequence of code signing, and of how Apple manages it through the App Store and on the device.

After the overview, which will give you some context for the whole process, I revert back to the natural order of things and start with getting your application to run on your iPhone during development.

How the Process Works

It's very important to keep clear that there are *two* different processes that you'll have to go through: one for development and one for distribution. Both of these processes produce different, but similarly named certificates and profiles, and you need to pay attention to keep them straight. I start with the *distribution* process — how you get your app to run on *other people's iPhones*. Then I go back and talk about the *development* process — how to get your app running on *your iPhone* during development.

The Distribution process

Before you can build a version of your application that will actually run on your users' iPhones, Apple insists that you have the following:

- ✔ **A Distribution Certificate:** An electronic document that associates a *digital identity* (which it creates) with other information that identifies you, including a name, e-mail address, or business that you have provided. The Distribution Certificate is placed on your *keychain* — that place on your Mac that securely stores passwords, keys, certificates, and notes for users.

- ✔ **A Distribution Provisioning Profile:** These profiles are code elements that Xcode builds into your application, creating a kind of "code fingerprint" that acts as a unique *digital signature*.

After you build your application for distribution, you then send it to Apple for approval and distribution. Apple verifies the signature to be sure that the code came from a registered developer (you) and hasn't been corrupted. Apple then adds its own digital signature to your signed application. The iPhone OS will only run applications that have that digital signature. Doing it this way ensures iPhone owners that the applications they download from iTunes have been written by registered developers and haven't been altered since they were created.

To install your distribution-ready application on a device, you can also create an *Ad Hoc Provisioning Profile,* which allows you to actually have your application used on up to 100 devices.

Although the system for getting apps on other people's iPhones works pretty well, leaving aside the fact that Apple essentially has veto rights on every application that comes its way, there are some significant consequences for developers. In this system, there really is no mechanism for testing your application on the device it's going to run on:

- ✔ You can't run your app on an actual device until it's been code-signed by Apple, *but* Apple is hardly going to code-sign something that may not be working correctly.

- ✔ Even if Apple did sign an app that hadn't yet run on an iPhone, that would mean an additional hassle: Every time you recompiled, you'd have to upload the app to the App Store again — *and* have it code-signed again because you had changed it, *and* then download it to your device.

Bit of a Catch-22 here.

The Development process

To deal with this problem, Apple has developed a process in which you can create a *Development Certificate* (as opposed to a Distribution Certificate that I explain in the preceding section) and a *Development Provisioning Profile* (as opposed to a Distribution Provisioning Profile that I also explain in the preceding section). It's easy to get these confused — the key words are Distribution and Development. With these items in hand, you can run your application on a *specific* device.

The Development Provisioning Profile is a collection of your App ID, Apple device UDID (a Unique Device Identifier for each iPhone), and iPhone Development Certificate (belonging to a specific developer). This Profile must be installed on each device on which you want to run your application code. (You find out how that is done later in this chapter.) Devices specified within the Development Provisioning Profile can be used for testing only by developers whose iPhone Development Certificates are included in the Provisioning Profile. A single device can also contain multiple provisioning profiles.

It's important to realize that a development provisioning profile (as opposed to a distribution one) *is tied to a device and a developer.*

Even with your provisioning profile(s) in place, when you compile your program, Xcode will build and sign (create the required signature for) your application *only* if it finds one of those Development Certificates in your keychain. Then when you install a signed application on your provisioned device, the iPhone OS verifies the signature to make sure that (a) the application was signed and (b) the application has not been altered since it was signed. If the signature is not valid or if you didn't sign the code, the iPhone OS won't let the application run.

This means that each Development Provisioning Profile is also tied to a particular Development Certificate.

And to make sure that the message has really gotten across:

A Development Provisioning Profile is tied to a *specific device* and a *specific Development Certificate.*

Your application, during development, must be tied to a specific *Development Provisioning Profile* (which is easily changeable).

The process you're about to go through is akin to filling out taxes: You have to follow the rules, or there can be some dire consequences. But if you do follow the rules, everything works out, and you don't have to worry about it again. (Until it's time to develop the next app, of course.)

While this is definitely not my favorite part of iPhone software development, I've made peace with it, and so should you. Now I go back to the natural order of things and start by explaining the process of getting your device ready for development. I'm happy to give you an overview of the process, but it will be up to you to go through it step by step on your own. Although Apple documents the steps very well, do keep in mind that you really have to carry them out in exactly the way Apple tells you. There are no shortcuts! But if you do it the way it prescribes, you'll be up and running on a real device very quickly.

With your app up and running, it's time for the next step: getting your creation ready for distribution. (I find that process to be somewhat easier.) Finally, you'll definitely want to find out how to get your application into the App Store. I aim to please, so I spell out those steps as well. After that, all you have to do is sit back and wait for fame and fortune to come your way.

This is the way things looked when I was writing this book. What you see when you go through this process yourself may be slightly different from what you see here. Don't panic. It's because Apple changes things from time to time.

Provisioning Your Device for Development

Until just recently, getting your application to run on the iPhone during development was a really painful process. In fact, I had written a 30-page chapter on it, detailing step after painful step. Then, lo and behold, right when I had put the finishing touches on my *magnum opus,* Apple changed the process and actually made it much easier. In fact, the process is now so easy that there's no real need for me to linger over the details. (Okay, I have some mixed feelings about that — but they're mostly *relief.*)

Here's the drill:

1. **Go to the iPhone Dev Center Web site at**

   ```
   http://developer.apple.com/iphone
   ```

   You should see the iPhone Developer Program section on the right side of the Web page (as shown in Figure 12-1). (Well, you should if you're a registered developer. You did take care of that, right? If not, look back at Chapter 3 for more on how to register.)

2. **Click the** iPhone Developer Program section's **iPhone Provisioning Portal button.**

   The iPhone Provisioning Portal screen appears, as shown in Figure 12-2.

3. **Assuming that you're either a Team Admin or Team Agent, or are enrolled in the Developer Program as an individual, use either the Development Provisioning Assistant to create and install a Provisioning Profile and iPhone Development Certificate, or allow Xcode to do it for you, as shown in the next section.**

   You need these profiles and certificates in order to build and install applications on the iPhone. But you knew that.

You've already identified yourself to Apple as one of two types of developers:

✔ If you're enrolled in the Developer Program as an **individual**, you're considered a Team Agent with all the rights and responsibilities.

✔ If you're part of a **company**, you've set up a team already. If not, click the Visit the Member Center Now link in the center of the screen to get more info about setting up a team and who needs to do what when.

This screen changes on a regular basis, so don't be surprised if it looks different when you visit it.

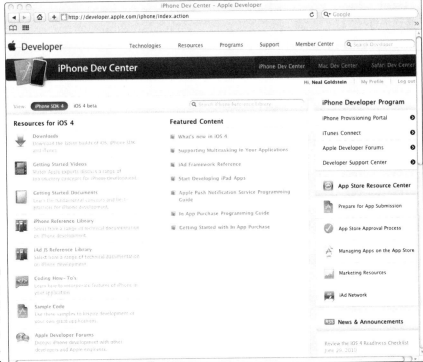

Figure 12-1:
The
gateway to
the Program
Portal.

Planning for your Development Provisioning Profile and iPhone Development Certificate

The first thing you need to do is generate a Development Certificate. The How-To's video, "Obtaining Your Certificate," in the right column of the Provisioning Portal does a stellar job of explaining how to do that.

After you do that, you have three choices:

1. Use the Development Provisioning Assistant.

2. Let Xcode create a provisioning profile for you.

3. Do it all on your own. (You'll have to explore this option on your own.)

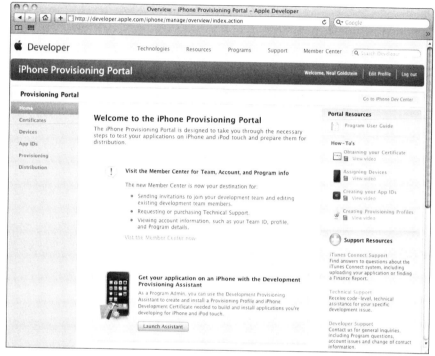

Figure 12-2:
Behold
the iPhone
Developer
Program
Portal.

Using the Development Provisioning Assistant

The first choice is to click the Launch Assistant button and simply go through the process.

As I mention earlier in the chapter, to run an application you're developing for iPhone, you must have a Provisioning Profile installed on the device on which you're running your app, as well as a Development Certificate on your Mac. The whole point of the Development Provisioning Assistant is to guide you through the steps to create and install your Development Provisioning Profile and iPhone Development Certificate.

Here's what the Development Provisioning Assistant has you do:

1. **Choose an App ID.**

   An App ID is a unique identifier that is one part of your Development Provisioning Profile.

Using the Assistant creates an App ID that *cannot* be used with the Apple Push Notification service, nor can it be used with In App Purchase or the Game Center. If you previously created an App ID that can be used with the Apple Push Notification service or for In App Purchase, you *can't* use the Assistant to create a Development Provisioning Profile. This is not a big deal; you just have to follow the steps the Assistant follows on your own.

2. **Choose an Apple Device.**

   Development provisioning is also about the device, so you have to specify which particular device you're going to use. You do that by providing the Assistant with the device's Unique Device Identifier (UDID), which the Assistant shows you how to locate by using Xcode.

3. **Provide your Development Certificate.**

   Because all applications must be signed by a valid certificate before they can run an Apple device, you should have created one at this point. If not, the How-To's video, "Obtaining Your Certificate," in the right column of the Developer Portal does a stellar job of explaining how to do that.

4. **Name your Provisioning Profile.**

   A Provisioning Profile pulls together your App ID (Step 1), Apple device UDID (Step 2), and iPhone Development Certificate (Step 3). The assistant steps you though downloading the profile and handing it over to Xcode, which installs it on your device.

Letting Xcode create a provisioning profile for you

You also have a second choice. After you create your Development Certificate, just go back to programming. Yep, Xcode (well, Xcode version 3.2.3 and newer, to be precise) will actually auto-provision your device for you. It will

1. Create an App ID for you with the name "AppStore ID."

2. Create a provisioning profile for your Team Provisioning Profile.

3. Automatically download the profile to your device.

All you need to do is plug in your device.

To start this process, here's what you do:

1. **Choose Window⇨Organizer from Xcode's main menu to open the Organizer window; then plug in your device.**

   You can see the result of this action in Figure 12-3.

2. **Click the Use For Development button.**

   Xcode will ask you for your iPhone Provisioning Portal logon (the same Apple ID you use to log in to the iPhone Dev Center). You can see Xcode's polite prompting in Figure 12-4.

3. **Supply the needed username and password and then click Log In.**

   You end up in (or back in) your Project window.

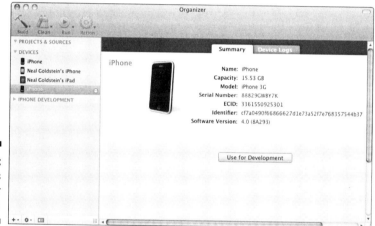

Figure 12-3:
Use this
device for
development.

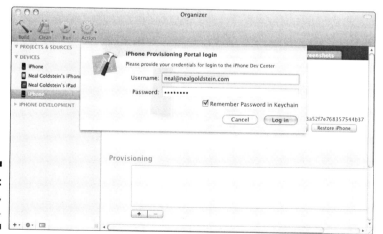

Figure 12-4:
Log in,
please.

4. In the Project window, choose iPhone Device as the active SDK in the Overview menu in the upper-left corner, as shown in Figure 12-5.

You can then build your application and have it installed on the provisioned device. You can see that in Figure 12-5, I also have Distribution as one of my active configuration options. Not to worry; you'll be there soon.

When you build and run your app, Figure 12-6 shows the (somewhat alarming) message you get. To get beyond this roadblock, just click Install and Run and you are there.

Just as with the Development Provisioning Assistant, the App ID that will be created *cannot* be used with the Apple Push Notification service, In App Purchase, or Game Center. If you previously created an App ID already that can be used with the Apple Push Notification service or for In App Purchase, you *can't* use the Assistant to create a Development Provisioning Profile. This is not a big deal; you just have to follow the steps the Assistant follows on your own.

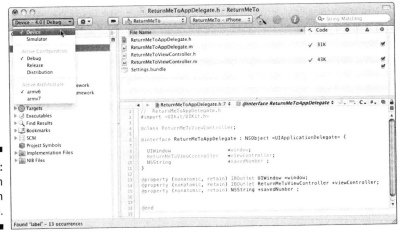

Figure 12-5:
Ready to run
your app on
the iPhone.

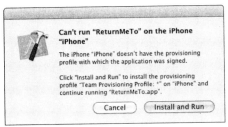

Figure 12-6:
Xcode will
install and
run your
profile for
you.

Provisioning Your Application for the App Store

Although there's no dedicated assistant to help you provision your application for the App Store, that process is actually a little easier — which may be why Apple didn't bother coming up with an assistant for it. You start at the Developer Portal (refer to Figure 12-2), but this time you select Distribution from the menu on the left side of the page. Doing so takes you to the screen shown in Figure 12-7 — an overview of the process, as well as links that take you where you need to go when you click them.

You actually jump through some of the very same hoops you did when you provisioned your device for development — except this time you're going after a distribution certificate.

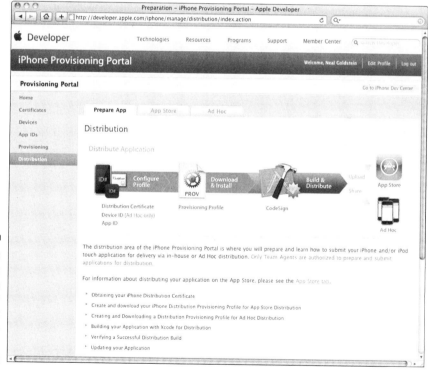

Figure 12-7:
Getting your app ready for distribution: You are here.

Here's what the overview highlights:

1. Obtaining your iPhone Distribution Certificate.

   To distribute your iPhone OS application, a Team Agent has to create an iPhone Distribution Certificate. This works much like the Development Certificate, except only the Team Agent (or whoever is enrolled as an Individual developer) can get one. Clicking the Obtaining Your iPhone Distribution Certificate link (shown in Figure 12-7) leads you through the process.

2. Create your iPhone Distribution Provisioning Profile for App Store Distribution.

   To build your application successfully with Xcode for distribution via the App Store, first you have to create and download an App Store Distribution Provisioning Profile — which is (lest we forget) *different* from the Development Provisioning Profiles I talk about in the preceding section.

   Clicking the Create and Download Your iPhone Distribution Provisioning Profile link (again, shown in Figure 12-7) leads you through the process.

   Apple will accept an application only when it's built with an App Store Distribution Provisioning Profile.

   If you had Xcode create your development profile for you, you can use the App ID it created here.

3. When you're done creating the Distribution Provisioning Profile, download it and drag it into Xcode on the Dock.

   That loads your Distribution Profile into Xcode, and you're ready to build an app you can distribute for use on actual iPhones.

4. (Optional) You can also create and download a Distribution Provisioning Profile for Ad Hoc Distribution.

   Going the Ad Hoc Distribution route enables you to distribute your application to up to 100 users without going through the App Store. Clicking the Creating and Downloading a Distribution Provisioning Profile for Ad Hoc Distribution link (refer yet again to Figure 12-7) leads you through the process. (Ad Hoc Distribution is beyond the scope of this book, but the iPhone Developer Program Portal has more info about this option.)

5. Build your application with Xcode for distribution.

   After you download the distribution profile, you can build your application for distribution — instead of just building it for testing purposes, which is what you've been doing so far. It's a well-documented process that you start by clicking the Building Your Application with Xcode for Distribution link (refer to Figure 12-7).

6. Verify that the build worked.

Click the Verifying a Successful Distribution Build link (refer to Figure 12-7) to get the verification process started. In this case, I find that there are some things missing in the heretofore well-explained step-by-step documentation, so I'll help you along.

If you check the handy documentation that's part of the Verifying a Successful Distribution Build link, it tells you to open the Build Log detail view and confirm the presence of the `embedded.mobileprovision` file. In Chapter 4, I show you how to keep the Build Results window open, but if you haven't been doing that, choose Build⇒Build Results.

Depending on the way the way the Build Results window is configured, you may see a window showing only the end result of your build, as in Figure 12-8, which shows that, yes, your build was in fact successful. You can see that in Figure 12-8, Errors & Warnings Only was selected in the drop-down menu in the scope bar. To get the actual log of the process, you have to change Errors & Warnings Only to All Messages in the drop-down menu in the scope bar, as I have in Figure 12-9.

Figure 12-8: And where's the transcript?

Figure 12-9: The build log revealed.

7. At this point, you'd be wise to do a couple of prudent checks:

- Verify that your application was signed by your iPhone Certificate. To do that, select the last line in the build log — the one that starts with `CodeSign`. Then click the icon at the end of the line, as I have in Figure 12-10.

  In Figure 12-11, you can see that it was signed by my iPhone Certificate. (Okay, you may need a magnifying glass, but trust me it's there, and make sure that yours is, too.) It should look something like this

  ```
  CodeSign build/Distribution-iphoneos/ReturnMeTo.app
  cd "/Users/neal/Desktop/ReturnMeTo"
  ```

- Verify that the `embedded.mobileprovision` is there and is located in the Distribution build directory and is not located in a Debug or Release build directory.

  To do that, search for `embedded.mobileprovision` in the Search field in the upper-right corner of the Build Results window, as I did in Figure 12-12. You see two matches. I chose the second one, and again clicked the icon at the end of the line to see more. I can see that it's there, and the directory it's building to is Distribution-iphoneos.

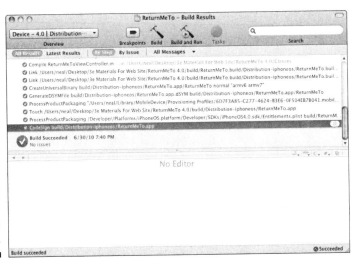

Figure 12-10: Getting more information from the Build log.

Figure 12-11:
It has been
signed by
my iPhone
Certificate.

Figure 12-12:
embedded.
mobileprovi-
sion is there
and building
to the right
directory.

When you've done this elaborate (but necessary) song and dance, you're
ready to rock 'n roll. You can go to iTunes Connect — your entryway to the
App Store. This is where the *real* fun starts.

iTunes Connect

Welcome to the world of forms, policies, and procedures. All of these are well documented by Apple, so I won't bore you (and more important me) by rewriting the Apple instructions.

There's also another reason for you to dive into the Apple instructions yourself. Remember Rule 1 — there are no shortcuts, so you need to follow the process exactly as Apple has documented it.

The exact details of the process, and all the rules, can be found in the Developer Guide. The link to download it is found at the bottom of the main iTunes Connect page in Figure 12-13. Follow it religiously (okay, enough already)

Although the rules may seem arbitrary, I'm going to cut the folks at Apple some slack here — given the number of apps they have to approve, there'd be absolute chaos if they didn't have a strict procedure in place.

But what I *can* do for you is give you an overview of the process so that what you have to do makes sense in the overall context. I also share with you some of the things I've learned in submitting my eight (and maybe even more by the time you read this) apps to the store.

Your portal into the world of fame, fortune, and forms is iTunes Connect. iTunes Connect is a group of Web-based tools that enables developers to submit and manage their applications for sale via the App Store. It's actually the very same set of tools that the other content providers — the music and video types — use to get their content into iTunes. In iTunes Connect, you can check on your contracts, manage users, and submit your application with all its supporting documentation — the *metadata*, as Apple calls it — to the App Store. This is also where you get financial reports and daily/weekly sales trend data (yea!).

You get here by clicking the iTunes Connect link in the iPhone Developer Program section of the iPhone Dev Center page. (Refer to Figure 12-1.) Before you can do anything, you're asked to review and accept the iTunes Distribution Terms & Conditions. After taking care of that chore, you land on the iTunes Connect page shown in Figure 12-13.

When you want to add an application to the App Store or manage what you already have there, the iTunes Connect main page is your control panel for getting that done.

You'll primarily be using three specific sections of the iTunes Connect page — the Manage Users section; the Contract, Tax & Banking Information section; and the Manage Your Applications section — so I take the time to explain those bits in greater detail in the next few pages.

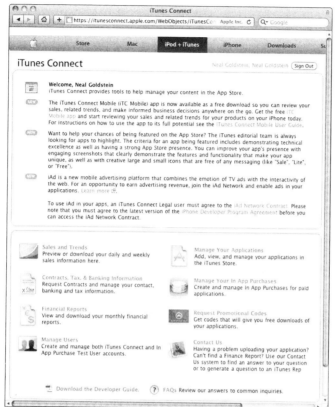

Figure 12-13:
The iTunes
Connect
main page.

Manage Users

Users in this context means you and your fellow team members, not any future potential users of your app. Click the Manage Users link to find out what tools are available for managing how you and your team communicate about what's what with your app. When creating and editing an iTunes Connect user account, you can define user roles and *notifications* — the type of e-mails your fellow team members will receive regarding the main iTunes Connect account. When setting up accounts, keep in mind that you have four distinct user roles to choose from: Admin, Legal, Finance, and Technical.

Contract, Tax & Banking Information

After you set up your various user accounts, proceed to the Contracts, Tax & Banking module. In this module, you need to complete the paid application agreements and provide financial information relating to payment and tax withholdings from the sale of your apps.

If you plan on selling your application, you need to have your paid commercial agreement in place and signed before your application can be posted to the App Store.

If your application is free, you've already entered into the freeware distribution agreement by being accepted into the iPhone Developer Program; however, there is still a contract setup that free application contracts need to go through before your application will go live in the App Store.

I'm not going to charge for the ReturnMeTo application, but just like with anything else at Apple, contract approval can take a while, so you should probably fill out the contract information just to get it out of the way.

If you're going to charge for your application, you have to provide even more information. Most of it is pretty straightforward, except for some of the banking information, which you do need to have available. To change some of the information after you enter it, you have to e-mail iTunes technical support. It behooves you to get it right the first time.

Here's what I'm talking about:

- ✔ **Bank name**
- ✔ **Bank address**
- ✔ **Account number**
- ✔ **Branch/Branch ID**
- ✔ **ABA/Routing Transit Number or SWIFT Code:** What this number is will depend upon where your bank is located. For United States banks, this number is the first nine digits of that long number at the bottom of your checks that also contains the account number. If you aren't sure what the routing number is, contact your bank. For non-U.S. banks, you may have to enter the SWIFT Code instead. You have to get that from your bank. The process also provides a look-up function to help you out.

Take it from me: It's far easier if you have all bits and pieces together *before you start the actual upload process,* instead of having to scramble at 3 a.m. to find some obscure piece of information you need. (The Bank SWIFT Code was the one that got me.)

Uploading your information

After you set the wheels in motion, you can then go back to the iTunes Connect main page and upload your data. At this point, you can start the application-upload process by clicking the Manage Your Applications link on the iTunes Connect main page. Make sure that you've dealt with the requisite Contracts, Tax & Banking Information.

The first time you enter the Manage Your Applications page in iTunes Connect, you see a blank page. After you uploaded your first binary, you can see your application(s) listed here.

To begin adding an application, click the Add New Application button. You will be taken through a series of pages that has you add the information required to get your app into the App Store.

But before you actually start the process, you have to have some things ready. I cover this in the next section.

Knowing what info to have in hand to get your app into the App Store

To start with, there's a link on the iPhone Dev Center page, under News and Information labeled: Tips on Submitting Your App to the App Store.

This page has information on Keywords, Assigning a Rating for Your App, and some other tips. Read it!

Apple is very strict about some things, and I speak from firsthand experience.

The first time I submitted ReturnMeTo to the App Store, I received a polite, but firm e-mail rejecting my application because my program icon used an iPhone. (You can see that icon and the rejection letters in the first edition of this book.) When I resubmitted my app, it was rejected a second time because I used an iPhone image as the image in the view you created in Chapter 5. (You can also see that image and the rejection letter in the first edition of this book.)

At the time of my first submissions, I really didn't think this whole image stuff was a big deal — hey, this was early on in the process — but what a difference a few rejections can make. Take it from me: Apple requires you to follow its guidelines. Period.

I fixed both of those issues, and you can now find ReturnMeTo in the App Store.

So how do you get your app into the App Store? Actually, the Uploading Your Application to the App Store part is pretty easy. The hard part is collecting all the little bits of information you'll need to enter into all the text fields in the upload page.

Here's an overview of the kind of information you'll need (for more information, download the Developer Guide, using the link at the bottom of the iTunes Connect page — see Figure 12-13):

✔ **Metadata:** The ever-present data about data. Here's what Apple wants from you:

- *Application Name:* The name must conform to guidelines for using Apple trademarks and copyrights. They take this very seriously, as evidenced by Apple sending a cease-and-desist order to my ISP when I tried (innocently) to use iPhoneDev411 as my domain name. (A word to the wise: Don't mess with Apple.)

- *Application Description:* When you go through the process of uploading your data, the field you have to paste this into says you're limited to 4,000 characters. Apple suggests no more than 580 characters so customers can view your entire iTunes Connect Application Description without clicking the More button in the App Store.

  This is what users will see when they click your app in the App Store, so it's important that this description be well written and that it point out all your app's key features.

  Don't include HTML tags; they will be stripped out when the data is uploaded. Only line breaks are respected.

- *Device:* Basically, we're talking iPhone and/or iPod touch.

- *Primary Category:* There will be a drop-down menu from which to choose the primary category for your app. There are about 20 choices, ranging from Business to Games to Social Networking to Travel to Utility.

- *Secondary Category:* (Optional) Same categories as the primary category.

- *Rating Information:* Later, you'll be asked to provide additional information describing the content. You'll see things like Cartoon or Fantasy Violence, Simulated Gambling, Mature/Suggestive Themes, and so on. For each type of content, you'll need to describe the level of frequency for that content — None, Infrequent/Mild, Frequent/Intense. This allows you to set your rating for your application for the purpose of parental controls on the iPhone App Store. Apple has strict rules stating that an app must not contain any obscene, pornographic, or offensive content. Oh and by the way, it's entirely up to Apple what is to be considered offensive or inappropriate.

- *Copyright:* I use this line:

  © Copyright Neal Goldstein 2010. All rights reserved.

  You can get the copyright symbol, in Word at least, by choosing Insert⇨Symbol and then selecting the copyright symbol. If you have any questions about copyright registration, talk to your lawyer or check out www.copyright.gov.

- *Version Number:* People usually start with 1.0. Then as you get more and more suggestions and "constructive criticism," you can move on to 1.1, and someday even version 2.0.

- *SKU Number:* A Stock Keeping Unit (SKU), any alphanumeric sequence of letters and numbers that is used to uniquely identify your application in the system. (Be warned — this is not editable after you submit it.)

- *Keywords:* Keywords that describe your app. These are matched to App Store searches. Spend some time on this one. Keywords can be changed only when you submit a new binary, or if the application status is Rejected, or Developer Rejected. Keywords must be related to your application content and cannot contain offensive or trademarked terms. You may not use other app names or company names as keywords. Keywords can be single words or phrases, and the text field is limited to 100 characters. Your App Name and Company are already searchable, so you don't need to include them in your list.

- *Support URL for the Company:* You need a support URL, which basically means you need a Web site, which isn't that hard. If you don't have a Web site yet and don't know how to build one, just go to your friendly ISP, find a domain name, get a package designed for folks who don't know HTML, and build yourself a Web site. Later on, you can get a hold of David Crowder's *Building a Web Site For Dummies,* 3rd Edition, (Wiley) which can help you build a "more professional" site. There will be a link to your support URL on the application product page at the App Store, and this is the link users will click on if they need technical support from you or have a question about your app.

- *Support E-Mail Address:* (For use by Apple only) Likely, this address will be the one you used when you registered for the developer program.

- *Demo Account – Full Access:* Test accounts that the App Store reviewers can use to test your application. Include usernames, passwords, access codes, demo data, and so on. Make sure that the demo account works correctly. You'd hate to have your app rejected because you didn't pay attention to setting up a demo account correctly.

- *End User License Agreement:* (Optional) If you don't know what this is, don't worry. It's the legal document that spells out to the end users what they're agreeing to do in order to use your app. Fortunately, the iTunes Store has a standard one. By this time, I think it probably knows what it's doing — but you should read it anyway before you use it.

- *Availability Date:* When your application will be available for purchase.

- *Application Price:* Free is easier, but if you want to get paid, you have to select a price tier. The last time I tried it, you couldn't see the pricing matrix unless you had first selected one. To help you along, Tier 1 is US$0.99 and so on.

- *Localization:* Additional languages (besides English) for your metadata. You can have your text and images in Italian in all Italian-speaking stores, for example.

- *App Store Availability:* The territories in which you would like to make your application available. (The default is all countries iTunes supports.)

✓ **Artwork:** A picture is worth a thousand words, so the App Store gives you the opportunity to dazzle your app's potential users with some nice imagery:

- *iPhone/iPod touch Home Screen Icon:* Your built application must have a 57-x-57-pixel icon included for it, following the procedure I show you back in Chapter 7. This icon is what will be displayed on the iPod touch or iPhone home screen.

- *Large Application Icon:* This icon will be used to display your application on the App Storefront. It needs to meet the following requirements:

  512 x 512 pixels (flattened, square image)

  72 dpi

  JPEG or TIFF format

- *Primary Screenshot:* This shot will be used on your application product page in the App Store.

  Apple doesn't want you to include the iPhone status bar in your screenshot. The shot itself needs to meet these requirements:

  320 x 460 portrait (without status bar) minimum

  480 x 300 landscape (without status bar) minimum

  320 x 480 portrait (full screen)

  Up to four additional optional screenshots can be on the application product page. These may be resized to fit the space provided. Follow the same requirements listed in this bullet.

- *Additional Artwork:* (Optional) If you're really lucky — I mean *really* lucky (or that good) — you may be featured on the App Store. Apple will want "high-quality layered artwork with a title treatment for your application," which will then be used in small posters to feature your application on the App Store.

The wait begins.

Avoiding the App Store Rejection Slip

During his opening Worldwide Developer Conference (WWDC) keynote address in June of 2010, Apple CEO Steve Jobs mentioned that there are now over 225,000 applications available. He also said that 15,000 applications are submitted per week, and that 95 percent of all apps submitted are approved within one week.

That's a pretty high approval rate. But just to be sure, it wouldn't hurt to know what was up with the 5 percent that *did* get rejected. It turns out that the majority of those rejected were rejected for one of the following reasons:

✔ The app doesn't function as advertised by the developer.

✔ The app uses private API's.

✔ The app crashes.

Sounds reasonable, but in addition to the Big Three, there are a few other reasons why apps are rejected. Keep the following in mind so that Apple smiles benignly down on your app:

✔ **Use the same icon for the app (the bundle icon) and the App Store page icon.** Make sure that the 57-pixel bundle icon for your app is the same image as the 512-pixel version for your App Store page.

✔ **Icons must be different for *lite* and *pro* versions (such as free and paid versions).** Make sure that you use a different icon image for a lite version than the one you use for the pro version.

✔ **Don't use any part of an Apple image and certainly none of the company's trademark images or names.** Your app can't include any photos or illustrations of the iPhone, including icons that resemble the iPhone, or any other Apple products, including the Apple logo itself. (You can read about my experience with this in the section "Knowing what info to have in hand to get your app into the App Store," earlier in this chapter.)

Your app can't include the word *iPhone* in its title.

If there is any doubt in your mind, it pays to read Apple's posted Guidelines for Using Apple's Trademarks and Copyrights, which you can find at

```
www.apple.com/legal/trademark/guidelinesfor3rdparties.
     html
```

✔ **If you use any of Apple's user interface graphics, you must use them in the way they were intended.** For example, the blue + button should be used only to add an item to a list.

- ✔ **Don't infringe on other trademarks.** Your app's title, description, and content must not potentially infringe upon other non-Apple trademarks or product likenesses.

- ✔ **Keywords can get you in trouble.** Keyword terms must be related to your app's content. It should be obvious, but some developers do it: You can't use offensive terms. And it's a big no-no to refer to other apps, competitive or not.

- ✔ **Don't include pricing information in your app's description and release notes.** Your app's marketing text — the application description and release notes — should not include pricing information, mostly because it would cause confusion in other countries due to pricing differences.

- ✔ **Don't mention Steve Jobs.** Apple will reject any app that mentions Steve Jobs in any context.

- ✔ **Don't try to fool the ratings.** Apps are rated accordingly for the highest (meaning most adult) level of content that the user is able to access. If you hide it, they will find it.

Now What?

Wait some more. As of June 30, 2010, 85 percent of new apps, and 95 percent of app updates were approved with one week (and this was a busy time because developers were submitting apps updated to iOS).

Part IV
An Industrial-Strength Application

The 5th Wave   By Rich Tennant

"Other than this little glitch with the landscape view, I really love my iPhone."

In this part . . .

After the realization dawns that you have to sell a lot of 99-cent applications to afford a meal in London, you can start thinking about raising the bar. Maybe what you need to do is develop an industrial-strength application that you can charge real money for — something so good that people will actually pay the big bucks for it. In this part, I explain the design of an application that has big muscles: a context-driven user interface, lots of functionality, Web access, an annotated custom map, local notifications in both the foreground and background, and an application architecture that you can use to build your own version of The Next Great Thing.

Chapter 13

Designing Your Application

*R*eturnMeTo, the star of Part III, is a useful little application and a great way to find out about iPhone software development. Apple considers it a utility application — like the Weather application, but with a single view.

Utility applications can provide real value to the user and are also fun and easy to write — not a bad combination. The good news is that by now, if you've been following along with me in the book, you understand enough about the framework, its architecture, components, and control flow to figure out how to build a nice little utility application on your own.

Not that you're all set right now to do everything you'd like — far from it. The way the Weather application flips the view, for example, may seem a total mystery. But Apple goes out of its way to provide samples for many of the neater tricks and features out there, all in hopes of demystifying how they work. In Chapter 19, I give you an annotated tour of the samples. With the ReturnMeTo application under your belt, it'll be a lot easier to understand and use all the resources Apple provides to help you develop iPhone applications.

What Apple doesn't show you (and where there's a real opportunity to develop a killer app) is how to design and develop more complex applications. Now, "more complex" doesn't necessarily — and shouldn't — mean "more complex *to the user.*" The real challenge and opportunity are in creating complex applications that are as easy to use as simple ones.

The iPhone Advantage

Because of its ease of use and convenience, its awareness of your location, and its ability to connect seamlessly to the Internet from most places, the iPhone lets you develop a totally new kind of application — one that integrates seamlessly with what the user is doing when he or she is living in the real world (what a concept). It frees the user to take advantage of technology away from the tether of the desk or coffee shop, and skips the hunt for a place to spread out the hardware. I refer to such applications as *here-and-now* — apps that take advantage of technology to help you do a specific task with up-to-date information, wherever you are and whenever you like. These applications move the user to the zero degrees of separation I speak of in Chapter 1.

All these features inherent in iPhone applications enable you to add a depth to the user's experience that you usually don't find in laptop- or desktop-based applications — in effect, a third dimension. Not only does the use of the application on the iPhone become embedded in where the user is and what the user is doing, the reverse is also happening: *Where the user is and what the user is doing can be embedded in the application itself.* This mutual embedding further blurs the boundaries between technology and user, as well as between user and task. Finally, developers can achieve a goal that's been elusive for years — the seamless integration of technology into everyday life.

The why-bother-because-I-have-my-laptop crowd still has to wrestle with this level of technology, especially those who haven't grown up with it. They still look at an iPhone as a poor substitute for a laptop or desktop — well, okay, for certain tasks, that's true. But an iPhone application trumps the laptop or desktop big-time in two ways:

✔ The iPhone's compact portability lets you do stuff not easily done on a laptop or desktop — on-site and right now — as with the MobileTravel411 application you're about to find out about.

✔ The iPhone is integrated into the activity itself, creating a transparency that makes it as unobtrusive and undistracting as possible. This advantage — even more important than portability — is the result of *context-driven design.*

The key to designing a killer iPhone application is understanding that the iPhone is *not* a small, more portable version of a laptop computer. It's another animal altogether, and is therefore used entirely differently. So don't go out and design (say) the ultimate word-processing program for an iPhone. (Given the device's limitations, I'd rather use a laptop.) But for point-in-time, 30-second tasks that may provide valuable information — and in doing so make someone's life much easier — the iPhone can't be beat.

In this chapter, I take you through an overview of the design cycle of a more complex application (MobileTravel411) and the resulting user interface and program architecture. While there are at least half a dozen models for the process (I'm a recovering software-development methodologist myself), the one I go through here is pretty simple and is well suited for the iPhone to boot. Here goes:

1. Defining the problems

2. Designing the user experience

    a. Understanding the real-world context

    b. Understanding the device context

    c. Categorizing the problems and defining the solutions

3. Creating the program architecture

    a. A main view

    b. Content views

    c. View controllers

    d. Models

4. Writing the code

5. Doing it until you get it right

After taking you through all of that, I show you how to develop a subset (iPhoneTravel411) of the application.

Of course, the actual analysis, design, and programming (not to mention testing) process has a bit more to it than this — and the specification and design definitely involve more than what you see in these few pages. But from a process perspective, it's pretty close to the real thing. It does give you an idea of the questions you need to ask — and have answered — to develop an iPhone application.

A word of caution, though. Even though iPhone apps are smaller and much easier to get your head around than, say, a full-blown enterprise service-oriented architecture, they come equipped with a unique set of challenges. Between the iPhone platform limitations I talk about in Chapter 1 and the high expectations of iPhone users, you have your hands full.

Defining the Problems

Innovation is usually born of frustration, and the MobileTravel411 project was no exception. It just turns out that my frustration was linked to a trip to beautiful Venice instead of, say, the vacuum cleaner doing a terrible job of picking up cat hair.

My wife and I were going to arrive late at night, and instead of trying to get into Venice at that hour, we decided we'd stay at a hotel near the airport and then go into Venice the next day. We were going to meet some friends who were leaving the day after that, and we wanted to get a relatively early start so we could spend the day with them.

I was a little concerned about the logistics. I thought we would have to go back to the airport terminal and then get on a water bus or water taxi. Both the water taxi stand and the water bus stop are a distance from the terminal, and that meant more time and more trudging about. The water taxi was the fastest way, but very pricey (around $140 USD at the time). The water bus was much cheaper but more confusing — and only ran once an hour. It seemed like a major excursion.

My friends said, "Why not take a taxi or a bus?"

I said, "A bus to Venice — it's an island, for crying out loud."

Okay, it *is* an island, but there's a causeway running from the mainland to Piazzale Roma, where you can then get a water bus or water taxi — or meet your friends.

Although it's more romantic to arrive by sea, it's a lot easier by land. Having been to Venice a couple times before, and considering our time constraints, we opted for the land route.

Now, I'm sure that information was in a guidebook someplace, but it would have taken a lot of work to dig it out; most guidebooks focus on attractions. Also, guidebooks go out of date quickly; the one I had for Venice was already two years old. Of course, I could have used the Internet before I left home to find the information, but that can also be a real chore.

What I wanted was something that made it easier to travel — reducing all those hassles — getting to and from a strange airport, getting around the city, getting the best exchange rate, knowing how much I should tip in a restaurant — that sort of thing. (Not too much to ask, right?)

Don't get me wrong — I actually do a lot of research before I go someplace, and often I have that information handy already. But I end up with lots of paper because I usually don't take a laptop with me on vacation; even when I do, it's terribly inconvenient to have to take it out on a bus or in an airline terminal to find some information. And then there's the challenge of finding a Wi-Fi connection when you really need it.

The iPhone is the perfect device to solve all those problems, so I decided to develop an iPhone application. It became MobileTravel411.

Designing the User Experience

To meet my Venetian (and other travel) needs, I didn't need a lot of information at any one time. In fact, what I wanted was as little as possible — just the facts, ma'am — but I wanted it to be as current as possible. It doesn't help to have last year's train schedule.

To get the design ball of my application rolling, I started by thinking about what I wanted from the application — not necessarily the features, but what the experience of using the application should be like.

Understanding the real-world context

You can reach the goal of seamlessness and transparency that I describe in the previous section by following some very simple principles when you design the user experience — especially with respect to the user interface.

Become the champion of relevance

There are two aspects to this directive:

- ✔ Search and destroy anything that is not relevant to what the user is doing while he or she is using a particular part of your application.

- ✔ Include — and make easily accessible — everything a user needs when doing something supported by a particular part of your application.

You want to avoid distracting the user from what he or she is doing. The application should be integrated into the task, a natural part of the flow, and not something that causes a detour. Your goal is to supply the user with only the information that's applicable to the task at hand. If your users just want to get from an airport into a city, they couldn't care less that the city has a world-renowned underground or subway system if it doesn't come out to the airport.

Seconds count

At first, the "seconds count" admonition may appear to fall into the "blinding flash of the obvious" category — of *course* a user wants to accomplish a task as quickly as possible. If the user has to scroll through lots of menus or figure out how the application works, the app's value drops off exponentially with the amount of time it takes to get to where the user needs to be.

But there are also some subtleties to this issue. If the user can do things as quickly as possible, he or she is a lot less distracted from the task at hand — and *both* results are desirable. As with relevance, this goal requires a seamless and transparent application.

Combine these ideas and you get the principle of *Simply Connect:* You want to be able to connect easily — to a network, to the information you need, or to the task you want to do. For example, a friend of mine was telling me he uses his iPhone when watching TV so he can look up things in an online dictionary or Wikipedia. (He must watch a lot of Public TV.)

The quality of information has to be better than the alternative

What you get by using the application has to have more value than alternative ways of doing the same thing. I can find airport transportation in a guidebook, but it's not up-to-date. I can get foreign exchange information from a *bureau de change,* but unless I know the bank rate, I don't know whether I'm being ripped off. I can get restaurant information from a newspaper, but I don't know whether the restaurant has subsequently changed hours or is closed for vacation. If the application can consistently provide me with better, more up-to-date information, it's the kind of application that's tailor-made for a context-driven design.

The app has to be worth the real cost

By *real cost,* I don't mean just the time and effort of using the application — you need to include the amount you actually pay out. The real cost includes both the cost of the application and any costs you might incur by *using* the application. This can be a real issue for an application such as MobileTravel411, because international roaming charges can be exorbitant. That's why the app must have the designed-in capability to download the information it provides and then to update the info when you find a wireless connection.

Keep things localized

With the world growing even flatter (from a communications perspective, anyway) and the iPhone available in more than 80 countries, the potential market for an app is considerably larger than just the folks who happen to speak English. But having to use an app in a language that you may not be comfortable with doesn't make for transparency. This means that applications have to be *localized* — that is, all the information, the content, and even the text in dialogs need to be in the user's language of choice.

Paying particular attention to three iPhone features

Key to creating applications that go beyond the desktop and that take advantage of context-based design are three hardware features of the iPhone. There are others, of course, but you can expect to find one or more of the following features in a context-based application.

Knowing the location of the user

This enables you to further refine the context by including the actual physical location and adding that to the relevance filter. If you are in London, the application can "ask" the user if he or she wants to use London as a filter for relevant information.

Accessing the Internet

Accessing the Internet allows you to provide real-time, up-to-date information. In addition, it enables you to transcend the CPU and memory limitations of the iPhone by offloading processing and data storage out to a server in the clouds.

Tracking orientation and motion

While used extensively in games, or to enable a user to erase a picture or make a random song selection by shaking the device, the accelerometers have potential in other kinds of applications. I recently saw one that alerts a company if its employees have a major change in acceleration. The change could mean any number of things — maybe they were driving too fast and they stopped suddenly, or maybe someone just fell off a ladder and hit the ground.

I leave it to you to debate the ethics and morality of these kinds of applications. But they do provide some food for thought on other application possibilities, and they certainly do get you some interesting context information about the user at a point in time.

Incorporating the device context

You have to take into account not only the user context, but also the device context.

After all, the device is also a context for the user. He or she, based on individual experience, expects applications to behave in a certain way. As I explain in Chapter 1, this expectation provides another perspective on why staying consistent with the user-interface guidelines is so important.

In addition to the device being a context from a user perspective, it's also one from the developer's perspective. If you want to maximize the user experience, you have to take the following into account (I know I go through these in Chapter 1, but remembering them is critical):

- ✔ **Limited screen real estate:** Although scrolling is built in to an iPhone and is relatively easy to do, you should require as little scrolling as possible, especially on navigation pages, and especially on the main page.

- ✔ **Limitations of a touch-based interface:** Although the Multi-Touch interface is an iPhone feature, it brings with it limitations as well. Fingers aren't as precise as a mouse pointer, and user interface elements need to be large enough and spaced far enough apart so that the user's fingers can find their way around the interface comfortably. You also can do only so much with fingers. There are definitely fewer possibilities when using fingers than when using the combination of multi-button mouse and keyboard.

- ✔ **Limited computer power, memory, and battery life:** As an application designer for the iPhone, you have to keep these issues in mind. The iPhone OS is particularly unforgiving when it comes to memory usage. If you run out of memory, the iPhone OS will simply shut your app down.

- ✔ **Connection limitations:** There's always a possibility that the user may be out of range, or on a plane, or has decided not to pay exorbitant roaming fees, or is using an iPod touch, which doesn't have Internet access except via Wi-Fi. You need to account for that possibility in your application and preserve as much functionality as possible. This usually means allowing the user to download and use the current real-time information, where applicable.

Again, all of this is covered in the detail you need in Chapter 1 and throughout the book.

Some of these goals overlap, of course, and that's where the real challenges are.

Categorizing the problems and defining the solutions

Because the app's requirements — and common sense — precluded scrolling through lots of information to get to what I needed, I had to create a *hierarchy*, which is a way of ordering the information or functionality into groups that makes sense to the user. On desktop or laptop machines, features are often categorized by function, but given the way the iPhone is used (as I describe in Chapter 1), categorizing by *context* makes more sense. (If you're interested

in the function versus context discussion, check out my Web site at
www.nealgoldstein.com for more information.) So after I settled on the
information and functionality I needed when I was traveling, I grouped things
into the following contexts.

- ✔ **Getting and using money:** What is the country's currency (including
  denominations and coins), and what's the best way to exchange my
  currency for it? I want to understand the costs of using credit cards
  versus an ATM card, or exchanging at a *bureau de change.* I also want
  to be able to understand how the dreaded VAT (value-added tax) really
  works.

- ✔ **Getting to and from the airport:** What choices do I really have when
  it comes to things terminal? What are the costs, advantages, and
  disadvantages — and logistics — of each? Do I have to buy a ticket in
  advance? How do I find said ticket? What is the schedule?

- ✔ **Getting around the city:** I'm in the same kind of pickle as getting to and
  from the airport — what's available *and* best for a traveler's purposes?
  I once spent several days in Barcelona before I realized that there was a
  subway system.

- ✔ **Seeing what's happening right now in the city:** Guidebooks are fine for
  visiting the sights, and I had no need to re-create one on the iPhone. But
  I *would* like to know if there's anything special happening when I'm in
  some particular place at some particular time. Bastille Day in Paris can
  be fun if you know about the Bastille Day parade, and less of a hassle if
  you know you can't cross the Champs-Élysées for a few hours.

- ✔ **Knowing the practical day-to-day stuff:** How do I make calls into, out
  of, and within a given city? How much and when should I tip? What is
  acceptable and unacceptable behavior? For example, that it might be
  impolite to eat or drink something while walking down the street in
  Japan might not occur to someone from New York City.

- ✔ **Staying safe:** Being immediately informed of breaking news that could
  make things unsafe — large demonstrations or terrorist attacks, for
  example — would be high on my wish list. But even the more mundane
  things like the "dangerous" neighborhoods are important. What should
  you do in an emergency? A friend of ours had her passport stolen in
  Prague — at times like that, it would be nice to have the locations and
  phone numbers of embassies or consulates. This is stuff you hardly ever
  need, but when you need it, you need it right away.

- ✔ **What to do before I go:** In the past, I have forgotten to call my cell
  phone company before I leave home to get a roaming package and notify
  my credit card company that I'll be out of the country or far from home,
  so please, *please* don't decline my hotel charge in Vladivostok. I also
  want to be able to download all the information before I leave so I can
  look at it on the plane, or as part of my strategy for avoiding roaming
  charges or handling an unexpected lack of connections.

I also wanted to make the app easy to use for someone who isn't intimately involved with the design — and perhaps doesn't immediately share my take on the best way to organize the information. So, for each choice in the main window, I wanted to be able to add a few words of explanation about what each category contained.

The final user interface I came up with looks like the left side of Figure 13-1. On the right side of Figure 13-1 is the iPhoneTravel411 version — a subset of the MobileTravel411 application. Although it's a subset, there's enough there for me to introduce you to almost all the technology I used to create the real application.

The rest of the views follow the same general format — some general information with specific information about each category.

Figure 13-1:
Mobile
Travel411
and iPhone
Travel41.

Part of making the app easy to use involves giving users a way to set their preferences for how the app should work. On the left side of Figure 13-2, you can see the MobileTravel411 Settings view, on the right the iPhoneTravel411 subset. The two most important of these settings involves being able to work in a *stored data mode* — using previously stored data, instead of the current real-time version that requires Internet access. The idea is to download the information I need before I leave — and then only update it occasionally while I'm gone, so I can avoid data roaming charges and afford food other than ramen noodles on my trip. The other is to turn on and off monitoring of the user's location by using the Monitor Distance preference. In Chapter 18, I show you how to use Core Location to monitor a user's location and update them to where they are and remind them what to do as they are leaving home. The Monitor Distance preference will allow the user to turn monitoring off to save battery power or because he or she doesn't want to be reminded how far they are from their final destination.

Figure 13-2:
Use offline
data.

Creating the Program Architecture

Given the user interface I just described, how did I get from there to here?

Keeping things at a basic level — a level that will be familiar to those of you who worked through the ReturnMeTo application — the MobilTravel411 is made up of the following:

- ✔ **Models:** Model objects encapsulate the logic and (data) content of the application. There was no model object in ReturnMeTo per se (although I opine on model functionality in Chapter 8). For iPhoneTravel411, I show you how to design, implement, and use model objects.

- ✔ **Views:** Views present the user experience; you have to decide what information to display and how to display it. In the ReturnMeTo application, there was a single content view with controls as subviews. Now, with the iPhoneTravel411 app, there will be a main view and several content views.

- ✔ **Controllers:** Controllers manage the user experience. They connect the views that present the user experience with the models that provide the necessary content. In addition (as you'll see), controllers also manage the way the user navigates the application.

No big surprises here — especially because the MVC model (Model-View-Controller) is pretty much the basis for all iPhone application development projects. The trick here is coming up with just the right views, controllers, and model objects to get your project off the ground. Within the requirements I spelled out in the "Designing the User Experience" section earlier in the chapter, I came up with the elements highlighted in the next few sections.

I start with the views because they determine the functionality and information available in a given context — being at an airport and needing to get into the city for example.

A main view

This one was a no-brainer. The main view for MobileTravel411 (and for iPhoneTravel411) is a UITableView, no question about it. Table views are used a lot in iPhone applications to do two things:

- ✔ **Display hierarchal data:** Think of the iPod application, which gives you a list of albums, and if you select one, a list of songs.

- ✔ **Act as a table of contents (or for my purposes, contexts):** Now think of the Settings application, which gives you a list of applications that you can set preferences for. When you select one of those applications from the list, it takes you to a view that lists what preferences you are able to set, and a way to set them.

Content views

Content views are views that display the information the user wants — stuff like how many Zimbabwean dollars I can get for $2.75 or the weather in Aruba next week.

The views you create are based on the information and functionality that a user needs in that context. At the risk of oversimplification, I had to think about two types of views:

- ✓ **The How Many Zimbabwean Dollars Can I Get For $2.75 (US) View:** These kinds of views are characterized by the fact that user input is required to be able to deliver the goods. The user would have to enter $2.75 US and then request to see the equivalent amount in Zimbabwean Dollars. (According to MobileTravel411, the answer by the way, in mid-July 2010, is around 1,037.)

  Examples of this kind of view in the MobileTravel411 application are shown in Figures 13-4, 13-5 (left), and 13-6 (right). These are all `UIView`'s with controls, constructed the same way as you constructed the view in the ReturnMeTo app. You won't be creating any of them in the next few chapters, but after you create the application structure, you can easily add them on your own.

- ✓ **The Weather In Aruba Next Week View:** This view, and others like it, simply displays information that it gets for the view controller. These are views like Figures 13-3, 13-5 (right), and 13-6 (left).

These second types of view (and all the views in the iPhoneTravel411 application) are *Web views* for some good practical reasons. First and foremost, some of the views must be updated regularly. If I want the current price and schedule of the Heathrow Express, for example, data from last year (or even last week) may not help me. I also want the most current information about what's happening in the city I plan to visit.

Web views, in that context, are the perfect solution; they make it easy to access data from a central repository on the Internet. (Client-server is alive and well!)

As for other benefits of Web views, keep in mind that real-time access is not always necessary — sometimes it's perfectly fine to store some data on the iPhone. It turns out that Web views can easily display formatted data that's locally stored — very handy. (You took advantage of that fact when building the ReturnMeTo application.)

Finally, I use Web views because they can access Web sites. If users want more information on the Heathrow Express, they can get to the Heathrow Express Web site by simply touching a link.

Figure 13-4:
Currency
selector.

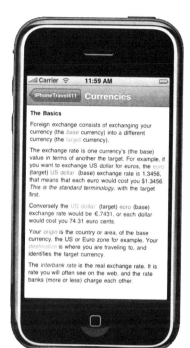

Figure 13-5:
Displaying
value in
dollars and
exchange-
rate
information.

Figure 13-6:
Weather
views.

View controllers

View controllers are responsible for not only providing the data for a view to display, but also responding to user input and navigation.

For each How Many Zimbabwean Dollars Can I Get For $2.75 (US) View, like the ones in Figure 13-4, on the left of 13-5, and on the right in Figure 13-6, I created a custom subclass of `UIViewController` to connect a control selected in the view (a button tapped, for example) with the model that has the logic and data to respond to the tap.

But for the views like those in Figure 13-3, the one on the right in Figure 13-5 and even the one on the left in Figure 13-6, for MobileTravel411, I actually designed a single subclass of `UIViewController` that connects the view to a model and can navigate from view to view for any context. That allows me to use the same view controller class to send to its Web view different content, from Getting To And From The Airport Data to Getting Around The City Data to Practical Day-To-Day Stuff Data. I did this by designing a view controller that can be initialized with the context information it needs when it's created (its model, the number of tabs, other things). While (fortunately for you) I won't be getting into the details of this, as I explain the view controllers you'll use in iPhoneTravel411, you'll notice the pattern.

Models

Although you could write a book on model design (in fact, I've written a couple, not to mention an Apple video — but that's another story), I want to concentrate on a couple things now to keep you focused. I elaborate more in Chapter 16.

When you begin to think about the models that you need for the application, you may think you've opened up a very large can of worms.

If you want to cover 160 cities or so — not unreasonable if you want your app to appeal to a broad, international audience — are you going to need a separate model object to deal with every city, airport, airport transportation option (trains, busses, taxis, and so on) and with every city transportation option (buses cars, subways, trains); and to deal with how to tip in every country, the way to make a phone call into, out of, and within every city, and so on? If so, you'll end up with thousands of classes (models, nib files, and so on). Definitely an unmanageable (not to mention resource-intensive) situation, and because there would also be a lot of redundancy in the code, maintenance would be a nightmare.

Fortunately, with some careful thought, you'll come to the same conclusion I did: No, not today, not tomorrow, not on your life.

No use reinventing the wheel

To show you the proper (and far less work-intensive) way to think about it, I want to review what the model objects need to do.

The models own the data and the application logic. In the MobileTravel411 application, for example, one model converts U.S. dollars to pounds (or any other currency) and vice versa. This kind of model is closely tied to the functionality of the view it supports. The How Many Zimbabwean Dollars Can I Get For $2.75 (US) View requires a model that can compute exchange rates, and here's where the real-world objects associated with object-oriented programming come into play. In the MobileTravel411 app, I have a Currency (model) object that knows how to compute exchange rates, and a VAT (value-added tax) object that does something similar. So for each view like those two, I create a model object.

When it comes to the Weather in Aruba Next Week View, though, the same approach holds, but not in an obvious way. You don't need a weather object here. The model object doesn't have to really care about the weather; all it really needs to do is have the logic to go out and get the data from a file, database, or server. For my purposes, a model object that gets weather information and a model object that gets information on the Heathrow schedule are pretty much the same. The logic for this object revolves around what the data is, how to access the data, and how this data may be connected to other data — the logic isn't about the *content* of the data. This means that, like the single view controller I mention earlier, I need only a single model class to support views like those in Figure 13-3, the one on the right in Figure 13-5 and even the one on the left in Figure 13-6.

So my path is clear: in MobileTravel411, I designed model objects in the same way I designed the view controllers — essentially creating a class that knows what data to include and where that data is to be found. (Programmers call such model objects *parameterized* models.) All you need to do is initialize the model with the context information it needs when the model is created.

All the model objects are of a subclass `NSObject`, because `NSObject` provides the basic interface to the runtime system. It already has methods for allocation, initialization, memory management, introspection (what class am I?), encoding and decoding (which makes it quite easy to save objects as "objects"), message dispatch, and a host of other equally obscure methods that I won't get into but are required for objects to be able to behave like they're expected to in an iPhone OS/Objective-C world.

Putting property lists to good use

To implement parameterized models and view controllers, you need something to provide the parameters. I used *property lists* — XML files, in other words — to take care of that because they're well suited for the job and (more important) support for them is built in to the iPhone frameworks.

Setting up property lists is a bit beyond the scope of this book, but in Chapter 16, the application structure I show you is conducive to using property lists to implement parameterized view controllers and model objects on your own.

Just as with the single view controller class, I don't get into the details of the single model class, but I explain its architecture in enough detail in Chapter 16 so that you'll see the pattern there as well.

But what I do in the iPhoneTravel411 application is actually have you create a model interface object and several model objects and view controllers to illustrate what you need to know about the model, view and controller relationship, how to access and display data stored locally or on a server, as well as how to simply display a Web site. That will be enough to keep you busy for a while.

User preferences, saving state, and localization

Using the application design I describe in the previous sections, adding these particular features is easy; I explain them as I work through the implementation in Chapters 14 and 15. Although I don't dig too deeply into localization in this book, I show you how to build your application so that you can easily include that handy feature in your app.

The Iterative Nature of the Process

If there's one thing I can guarantee about development, it's that *nobody gets it right the first time.* Although object-oriented design and development are in themselves fun intellectual exercises (at least for some of us), they also are very valuable. An object-oriented program is relatively easier to modify and extend, not just during initial development, but also over time from version to version. (Actually, "initial development" and "version updating" are both the same; they differ only by a period of rest and vacation between them.)

The design of my MobileTravel411 application evolved over time, as I learned the capabilities and intricacies of the platform and the impact of my design decisions. What I try to do in this chapter, and the ones following, is to help you avoid (at least most of) the blind alleys I stumbled down while developing my first application. So get ready for a stumble-free experience. On with the development!

Chapter 14

Setting the Table

*V*iews are the user's window into your application; they present the user experience on a silver platter, as it were. Their associated view controllers manage the user experience by providing the data displayed in the view, as well as by enabling user interaction.

In this chapter, you get a closer look at the iPhoneTravel411 *main view* — the view you see when you launch the application — as well as the view controller that enables it. As part of your tour of the main view, I show you how to use one of the most powerful features of the framework: table views. In the chapters that follow, I show you how to implement the views that you set up to deliver the content of your application — how to get from Point A to Point B, convert yen to yuan, or check on the weather in Anaheim, Azusa, or Cucamonga.

My running example here is the iPhoneTravel411 application described in Chapter 13. Space prohibits dotting every *i* and crossing every *t* in implementing the application, but I can show you how to use the technology you need to do the detailed work on your own. And even though I don't have the complete listings in this book, copies are available on my Web site. You'll work with the project in the folder named iPhoneTravel411 Chapter 16, which has the code for the finished application through Chapter 16. The iPhoneTravel411 Chapter 17 folder has the code for the complete application.

Working with Table Views

Table views are front and center in several applications that come with the iPhone out of the box; they play a major role in many of the more complex applications you can download from the App Store. (Obvious examples: Almost all the views in the Mail, iPod, and Contacts applications are table views.) Table views not only display data, but also serve as a way to navigate a hierarchy.

If you take a look at an application such as Mail or iPod, you'll find that table views present a scrollable list of *items* (or *rows* or *entries* — I use all three terms interchangeably) that may be divided into *sections*. A row can display text or images. So, when you select a row, you may be presented with another table view — or with some other view that may display a Web page or even some controls such as buttons and text fields. You can see an illustration of this diversity in Figure 14-1. Selecting Map on the left leads to a content view displaying a map of London and its environs — very handy after a long flight.

Figure 14-1:
A table and
Web view.

But while a table view is an instance of the class UITableView, each visible row of the table uses an UITableViewCell to draw its contents. Think of a *table view* as the object that creates and manages the table structure, and the *table-view cell* as being responsible for displaying the content of a single row of the table.

Creating the table view

Although powerful, table views are surprisingly easy to work with. To create a table, you need only do four — count 'em, four — things, in the following order:

1. **Create and format the view itself.**

   This includes specifying the table style and a few other parameters — most of which is done in Interface Builder.

2. **Specify the table-view configuration.**

   Not too complicated, actually. You let UITableView know how many sections you want, how many rows you want in each section, and what you want to call your section headers. You do that with the help of the numberOfSectionsInTableView: method, the tableView:numberO fRowsInSection: method, and the tableView:titleForHeaderIn Section: method, respectively.

3. **Supply the text (or graphic) for each row.**

   You return that from the implementation of the tableView:cellFor RowAtIndexPath: method. This message is sent for each visible row in the table view, and you return a table-view cell to display the text or graphic.

4. **Respond to a user selection of the row.**

   You use the tableView:didSelectRowAtIndexPath: method to take care of this task. In this method, you create a view controller and a new view. For example, when the user selects Map in Figure 14-1, this method is called, and then a Map controller and a Map view are created and displayed.

A UITableView object must have a *data source* and a *delegate*. The data source supplies the content for the table view, and the delegate manages the appearance and behavior of the table view. The data source adopts the UITableViewDataSource protocol, and the delegate adopts the UITableViewDelegate protocol — no surprises there. Of the preceding methods, only the tableView:didSelectRowAtIndexPath: is included in the UITableViewDelegate protocol. All the others I list earlier are included in the UITableViewDataSource protocol.

The data source and the delegate are often (but not necessarily) implemented in the same object — which is often a subclass of `UITableViewController`. I plan to use the `RootViewController` for my iPhone411Travel app.

Implementing these five (count 'em, five) methods — and taking Interface Builder for a spin or two, along with the same kind of initialization methods and the standard memory-management methods you used in the ReturnMeTo application, creates a table view that can respond to a selection made in the table.

Not bad.

Creating and formatting a grouped table view

Table views come in two basic styles. The default style is called *plain* and looks really unadorned — plain vanilla. It's a list: just one darn thing after another. You can index it, though, just as the table view in the Contacts application is indexed, so it can be a pretty powerful tool.

The other style is the *grouped* table view; unsurprisingly, it allows you to clump entries into various categories. In Figure 14-2, you can see a grouped table view on the left; the one on the right is a plain table view.

Grouped tables cannot have an index.

When you configure a grouped table view, you can also have header, footer, and section titles (a plain view can also have section headers and footers). I show you how to do section titles shortly.

To see how table views work, you of course need a project you can use to show them off. With that in mind, fire up Xcode and officially launch the iPhoneTravel411 project. (If you need a refresher on how to set up a project in Xcode, take another look at Chapter 4.) As you can see in Figure 14-3, you need to go with a Navigation-Based Application template.

When you select Choose, you see a standard Save sheet. I saved the project as iPhoneTravel411 in a folder on my desktop.

Your new project gets added to the Groups & Files list on the left side of the Xcode Project window. Next, take a look at what happens when you drill down in your project folder in the Groups & Files list until you end up selecting `RootViewController` (as shown in Figure 14-4). The main pane of the Xcode Project window reveals the fact that `RootViewController` is derived from a `UITableViewController`.

Figure 14-2:
Grouped
and plain
tables.

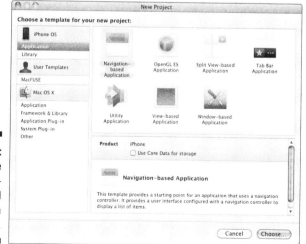

Figure 14-3:
The
Navigation-
Based
Application
template.

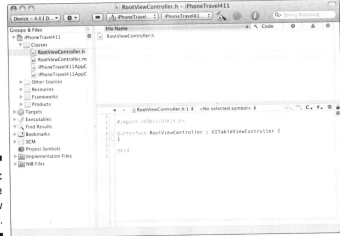

Figure 14-4:
The
RootView
Controller.

Inquisitive type that you are, you look up UITableViewController in the Documentation reference by right-clicking its entry and then choosing Find Selected Text in Documentation from the pop-up menu that appears. The Class reference tells you that UITableViewController conforms to the UITableViewDelegate and UITableViewDataSource protocols (and a few others) — the two protocols I said were necessary to implement table views. What luck. (Kidding. It's all intentional.)

Always on the lookout for more information, you continue down the Groups & Files list to open your project's Resources folder, where you double-click the RootViewController.xib file to launch Interface Builder. You are reassured to see a table view set up in front of you — admittedly, a plain table view rather than the grouped table view we want, but a table view nonetheless. To get the final duck in a row, choose Grouped from the Style drop-down menu in the Attributes Inspector, shown in Figure 14-5, to make the switch from plain to grouped. Be sure to save the file after you do this.

At this point, you can build and run this project; go for it. What you see in the Simulator is a table view — and if you try to scroll it, you get a "bounce scroll," where the view just bounces back up when you scroll it, but not much else. In fact, you won't even see it as a grouped view. What you do have is the basic framework, however, and now you can format it the way you like.

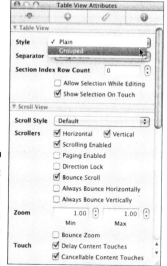

Figure 14-5:
Choosing
Grouped
in the
Attributes
Inspector.

Making UITableViewController work for you

The data source and the delegate for table views are often (but not necessarily) the same object — and that object is frequently a custom subclass of UITableViewController. For the iPhoneTravel411 project, the RootView Controller created by the Navigation-Based Application template is a subclass of UITableViewController — and the UITableView Controller has adopted the UITableViewDelegate and UITableView DataSource protocols. So you're free to implement those handy methods I mention in the "Creating the table view" section, earlier in the chapter. (Just remember that you need to implement them in RootViewController to make your table usable.) Start with the methods that format the table the way you like.

Adding sections

In a grouped table view, each group is referred to as a *section*.

In an indexed table, each indexed grouping of data is also called a section. For example, in the iPod application, all the albums beginning with "A" would be one section, those beginning with "B" another section, and so on. While having the same name, this is not the same thing as sections in a grouped table (which doesn't have an index).

The two methods you need to start things off are as follows:

```
numberOfSectionsInTableView:(UITableView *)tableView
tableView:(UITableView *)tableView
            numberOfRowsInSection:(NSInteger)section
```

Each of these methods returns an integer that tells the table view something — the number of sections and the number of rows in a given section, respectively.

In Listing 14-1, you can see the code that results in two sections with three rows in each section. These methods are already implemented for you by the Navigation-Based Application template in the `RootViewController.m` file. You just need to remove the existing code and replace it with what you see in Listing 14-1.

Listing 14-1: Modifying numberOfSectionsInTableView: and tableView:numberOfRowsInSection:

```
- (NSInteger)numberOfSectionsInTableView:
                                (UITableView *)tableView {

    return 2;
}

- (NSInteger)tableView:(UITableView *)tableView
            numberOfRowsInSection:(NSInteger)section {

    NSInteger rows;
    switch (section) {
      case 0:
        rows = 4;
        break;
      case 1:
        rows = 3;
        break;
      default:
        break;
    }
    return rows;
}
```

You implement `tableView:numberOfRowsInSection:` by using a simple `switch` statement:

```
switch (section) {
```

Keep in mind that the first section is zero, as is the first row.

Although that's as easy as it gets, it's not really the best way to do it. Read on.

In the interest of showing you how to implement a robust application, I'm going to use constants to represent the number of sections *and* the number of rows in each section. I put those constants in a file, `Constants.h`, which will eventually contain other constants. I do this for purely defensive reasons: Both of these values will be used often in this application (I know that because hindsight is 20-20), and declaring them as constants makes changing the number of rows and sections easy, and it also helps avoid hard-to-detect typing mistakes.

I show you some techniques here that make life much, much easier later. It means paying attention to some of the less glamorous application nuts-and-bolts functionality — can you say, "memory management"? — that may be annoying to implement along the way but that are *really* difficult to retrofit later. I want to head you away from the boulder-strewn paths that so many developers have gone down (me included), much to their later sorrow.

To implement the `Constants.h` file, you do the following:

1. **Choose File⇨New File from the Xcode main menu.**

   I recommend having the `Classes` folder selected in the Groups & Files list so the file will be placed in there.

2. **In the New File dialog that appears, choose Other from the listing on the left (under the Mac OS X heading) and then choose Empty File in the main pane, as shown in Figure 14-6.**

3. **In the new dialog that appears, name the file Constants.h, and then click Finish.**

   The new empty file is saved in the Classes folder, as shown in Figure 14-7.

With a new home for your constants all set up and waiting, all you have to do is add the constants you need so far. (Listing 14-2 shows you the constants you need to add to the `Constants.h` file.)

Listing 14-2: Adding to the Constants.h File

```
#define kSections        2
#define kSection1Rows     4
#define kSection2Rows     3
```

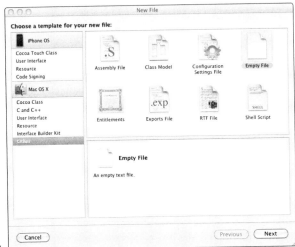

Figure 14-6:
Creating an
empty file.

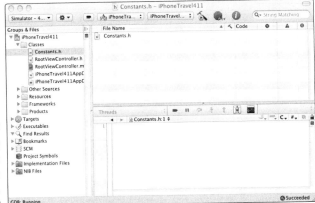

Figure 14-7:
The
Constants.h
file.

Having a `Constants.h` file in hand is great, but you have to let
`RootViewController.m` know that you plan to use it. To include
`Constants.h` in `RootViewController.m`, open `RootViewController.m`
in Xcode and add the following statement:

```
#import "Constants.h"
```

You can then use these constants in all the various methods used to create your table view, as shown in Listing 14-3.

Listing 14-3:   Sections and Rows Done Better

```
- (NSInteger)numberOfSectionsInTableView:(UITableView *)
                                      tableView {

  return kSections;
}

- (NSInteger)tableView:(UITableView *)tableView
            numberOfRowsInSection:(NSInteger)section {

  NSInteger rows;
  switch (section) {
    case 0:
      rows = kSection1Rows;
      break;
    case 1:
      rows = kSection2Rows;
      break;
    default:
      break;
  }
  return rows;
}
```

When you build and run this (provisional) app, you get what you see in Figure 14-8 — two sections, the first with four rows and the second with three.

Although using constants and a `switch` statement makes your program more extensible, it does require you to change the `switch` statement if you want to add or change the layout. An even better solution is to create an array in `awakeFromNib` that looks like this:

```
sectionsArray = [[NSMutableArray alloc]
                                  initWithCapacity:2];
[sectionsArray addObject:[[ NSNumber alloc]
                                  initWithInt:4]];
[sectionsArray addObject:[[ NSNumber alloc]
                                  initWithInt:3]];
```

Then you could use the array count `[sectionsArray count]` to return the number of sections, and index into the array for the number of rows in a section `[sectionsArray objectAtIndex:section]`.

Figure 14-8:
Now I have
sections.

Adding titles for the sections

With sections in place, you now need to title them so users know what the sections are for. Luckily for you, the UITableViewdataSource protocol has a handy method — titled, appropriately enough, the tableView:titleFor HeaderInSection: method — that enables you to add a title for each section. Listing 14-4 shows how to implement the method.

Listing 14-4:   Adding Section Titles

```
- (NSString *)tableView:(UITableView *)tableView
             titleForHeaderInSection:(NSInteger)section {

   NSString *title = nil;
   switch (section) {
      case 0:
         title = @"Welcome to London";
         break;
      case 1:
         title =  @"Getting there";
```

```
        break;
    default:
        break;
    }
    return title;
}
```

This (again) is a simple `switch` statement. For `case 0`, or the first section, you want the title to be `"Welcome to London"`, and for `case 1`, or the second section, you want the title to be `"Getting there"`.

Okay, this, too, was really easy — so you probably won't be surprised to find that it's *not* the best way to tackle the whole titling business. Another path not to take — in fact, a really *important* one not to take. Really Serious Application Developers insist on catering to the needs of an increasingly global audience, which means — paradoxically — that they have to *localize* their applications. In other words, an app must be created in such a way that it presents a different view to different, local audiences. The next section explains how this is done.

Localization

Localizing an application isn't difficult, just tedious. To localize your application, you create a folder in your application bundle (I'll get to that) for each language you want to support. Each folder has the application's translated resources.

In the Settings application for the iPhoneTravel411 app, you're going to set things up so the user can set the language — Spanish or Italian, for example — and the region format.

For example, if the user's language is Spanish, available regions range from Spain to Argentina to the United States and lots of places in between. When a localized application needs to load a resource (such as an image, property list, or nib), the application checks the user's language and region and looks for a localization folder that corresponds to the selected language and region. If it finds one, it loads the localized version of the resource rather than the *base* version — the one you're working in.

Showing you all the ins and outs of localizing your application is a bit too Byzantine for this book. But I *do* show you what you must do to make your app localizable when you're ready to tackle the chore on your own.

What you have to get right — right from the start — are the strings you use in your application that get presented to the user. (If the user has chosen Spanish as his or her language of choice, what's expected in the main view is now *Moneda*, not *Currency*.) You ensure that the users see what they're expecting by storing the strings you use in your application in a `strings` text file; this file contains a list of string pairs, each identified by a comment. You would create one of these files for each language you support.

Here's an example of what an entry in a strings file might look like for this application:

```
/*Airport choices */
"Getting there"  = "Getting there";
```

The values between the /* and the */ characters are just comments for the (human) translator you task with creating the right translation for the phrase — assuming, of course, that you're not fluent in the ten-or-so languages you'll probably want to include in your app, and therefore will need some translating help. You write such comments to provide some context — how that string is being used in the application.

Okay, this example has two strings — the one to the left of the equal sign is used as a key; the one to the right of the equal sign is the one displayed. In the example, both strings are the same — but in the strings file used for a Spanish speaker, here's what you'd see:

```
/*Airport choices */
"Getting there"  = "Cómo llegar";
```

Looking up such values in the table is handled by the NSLocalizedString macro in your code.

To show you how to use the macro, I take one of the section headings as an example. Rather than

```
title = Getting there;
```

I code it as follows:

```
title = NSLocalizedString(@"Getting there",
                          @"Airport choices");
```

As you can see, the macro has two inputs. The first is the string in your language, the second the general comment for the translator. At runtime, NSLocalizedString looks for a strings file named localizable. strings in the language that has been set: Spanish, for example. (A user would have done that by going to Settings, selecting General⇨Internation al⇨Language⇨Español.) If NSLocalizedString finds the strings file, it searches the file for a line that matches the first parameter. In this case, it would return "Cómo llegar," and that is what would be displayed as the section header. If the macro doesn't find the file or a specified string, it returns its first parameter — and the string will appear in the base language.

To create the `localizable.strings` file, you run a command-line program named `genstrings`, which searches your code files for the macro and places them all in a `localizable.strings` file (which it creates), ready for the (human) translator. `genstrings` is beyond the scope of this book, but it's well documented. When you're ready, I leave you to explore it on your own.

Okay, sure, it's really annoying to have to do this sort of thing as you write your code (yes, I know, *really, really* annoying). But that's not nearly as annoying as having to go back and *find and replace all the strings you want to localize* after the application is almost done. Take my word for it!

Listing 14-5 shows how to use the `NSLocalizedString` macros to create localizable section titles.

Listing 14-5: Adding Localizable Section Titles

```
- (NSString *)tableView:(UITableView *)tableView
            titleForHeaderInSection:(NSInteger)section {

  NSString *title = nil;
  switch (section) {
    case 0:
      title = NSLocalizedString(@"Welcome to London",
                                          @"City name");
      break;
    case 1:
      title = NSLocalizedString(@"Getting there",
                                       @"Airport choices");
      break;
    default:
      break;
  }
  return title;
}
```

Creating the row model

As all good iPhone app developers know, the Model-View-Controller (MVC) design pattern is the basis for the design of the framework you use to develop your applications. In this design pattern, each element (model, view, and controller) concentrates on the task at hand; it doesn't much care what the other elements are doing. For table views, that means the method that draws the content doesn't know what the content is — and the method that decides what to do when a selection is made in a particular row is equally ignorant of what the selection is. The important thing is to have a model object — one for each row — to hold and provide that information.

In this kind of situation, you usually want to deal with the model-object business by creating an array of models, one for each row. In our case, the model object will be a dictionary that holds the following three items:

- ✔ **The selection text:** Heathrow, for example.
- ✔ **The description text:** International airport, for example.
- ✔ **The view controller to be created when the user selects that row:** `AirportController`, for example.

You can see all three items illustrated in Figure 14-9.

In more complex applications, you could provide a dictionary *within* the dictionary and use it to provide the same kind of information for the next level in the hierarchy. The iPod application is an example: It presents you with a list of albums, and then when you select an album, it shows you a list of songs on that album.

The following code shows you how to create a single dictionary for a row. Later I show you how to create all the dictionaries and where all this code needs to go.

```
menuList = [[NSMutableArray alloc] init];

[menuList addObject:[NSMutableDictionary
                           dictionaryWithObjectsAndKeys:
     NSLocalizedString(@"Heathrow",
                     @"Heathrow Section"), kSelectKey,
     NSLocalizedString(@"International airport",
                     @"Heathrow Explain"), kDescriptKey,
     nil, kControllerKey, nil]];
```

Here's the blow-by-blow account:

1. **Create an array to hold the model for each row.**

    An `NSMutableArray` is a good choice here, because it allows you to easily insert and delete objects.

    In such an array, the position of the dictionary corresponds to the row it implements, that is, relative to row zero in the table and not taking into account the section.

Dictionary

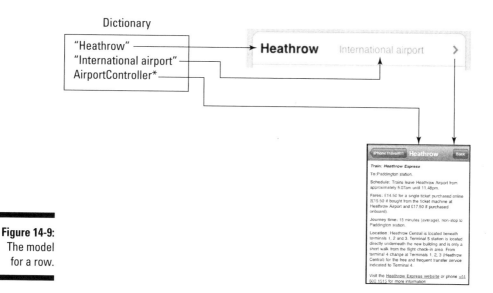

Figure 14-9:
The model
for a row.

2. **Create an `NSMutableDictionary` with three entries and the following keys:**

   - `kSelectKey`: The entry that corresponds to the main entry in the table view ("Heathrow," for example).

   - `kDescriptKey`: The entry that corresponds to the description in the table view ("International Airport," for example).

   - `kControllerKey`: This entry contains a pointer to a view controller that will display the Heathrow information. I'll create an entry for the controller, but not just yet; I just use `nil` for now. The first time the user selects a row, I'll create the view controller and save that value in here. That way, if the user selects that row again, the controller will simply be reused.

3. **Add the keys to the `Constants.h` file.**

   ```
   #define kSelectKey      @"selection"
   #define kDescriptKey    @"description"
   #define kControllerKey  @"viewController"
   ```

   The @ before each of the preceding strings tells the compiler that this is an `NSString`.

You'd use the same mechanism to get rid of these controllers if you were to ever get a low-memory warning. You'd simply go through each dictionary in the array and `release` every controller except the one that's currently active.

You'll want to create this array and all the dictionaries in an initialization method `awakeFromNib`, which you'll need to add to the `RootView Controller.m` file — you can see `awakeFromNib` in Listing 14-6. When the `RootViewController`'s nib file is loaded, the `awakeFromNib` message is sent to the `RootViewController` after all the objects in the nib file have been loaded and the `RootViewController`'s outlet instance variables have been set. Typically, the object that owns the nib file (File's Owner, in this case the `RootViewController`) implements `awakeFromNib` to do initialization that requires that outlet and Target-Action connections be set.

You could argue that you really should create a model class that creates this data-model array and get its data from a file or property list. For simplicity's sake, I do it in the `awakeFromNib` method for the iPhoneTravel411 app.

Listing 14-6: awakeFromNib

```
- (void)awakeFromNib {

  self.title = [[[NSBundle mainBundle] infoDictionary]
                               objectForKey:@"CFBundleName"];
  menuList = [[NSMutableArray alloc] init];
  [menuList addObject:[NSMutableDictionary
                               dictionaryWithObjectsAndKeys:
     NSLocalizedString(@"London", @"City Section"),
                                             kSelectKey,
     NSLocalizedString(@"What's happening",
                          @"City Explain"), kDescriptKey,
     nil, kControllerKey, nil]];
  [menuList addObject:[NSMutableDictionary
                               dictionaryWithObjectsAndKeys:
     NSLocalizedString(@"Map", @"Map Section"),
                                             kSelectKey,
     NSLocalizedString(@"Where you are",
                          @"Map Explain"), kDescriptKey,
     nil, kControllerKey, nil]];
  [menuList addObject:[NSMutableDictionary
                               dictionaryWithObjectsAndKeys:
     NSLocalizedString(@"Currency", @"Currency Section"),
                                             kSelectKey,
     NSLocalizedString(@"About foreign exchange",
                       @"Currency Explain"), kDescriptKey,
     nil, kControllerKey, nil]];
  [menuList addObject:[NSMutableDictionary
                               dictionaryWithObjectsAndKeys:
     NSLocalizedString(@"Weather", @"Weather Section"),
                                             kSelectKey,
     NSLocalizedString(@"Current conditions",
                       @"Weather  Explain"), kDescriptKey,
```

```
          nil, kControllerKey, nil]];
     [menuList addObject:[NSMutableDictionary
                              dictionaryWithObjectsAndKeys:
          NSLocalizedString(@"Heathrow", @"Heathrow Section"),
                                          kSelectKey,
          NSLocalizedString(@"International airport",
                          @"Heathrow Explain"), kDescriptKey,
          nil, kControllerKey, nil]];
     [menuList addObject:[NSMutableDictionary
                              dictionaryWithObjectsAndKeys:
          NSLocalizedString(@"Gatwick", @"Gatwick Section"),
                                          kSelectKey,
          NSLocalizedString(@"European flights",
                          @"Gatwick Explain"), kDescriptKey,
          nil, kControllerKey, nil]];
     [menuList addObject:[NSMutableDictionary
                              dictionaryWithObjectsAndKeys:
          NSLocalizedString(@"Stansted",
                          @"Stansted Section"), kSelectKey,
          NSLocalizedString(@"UK flights",
                          @"Stansted Explain"), kDescriptKey,
          nil, kControllerKey, nil]];
     self.destination = [[Destination alloc]
                                  initWithName:@"England"];

}
```

Going through the code, you can see that the first thing you do is get the application name from the bundle so you can use it as the main view title.

```
self.title = [[[NSBundle mainBundle] infoDictionary]
                          objectForKey:@"CFBundleName"];
```

"What bundle?" you ask. Well, when you build your iPhone application, Xcode packages it as a bundle — containing

- ✔ The application's executable code

- ✔ Any resources that the app has to use (for instance, the application icon, other images, and localized content)

- ✔ The info.plist, also known as the information property list, which defines key values for the application, such as bundle ID, version number, and display name

infoDictionary returns a dictionary that's constructed from the bundle's info.plist. CFBundleName is the key to the entry that contains the (localizable) application name on the home page. The title is what will be displayed in the navigation bar at the top of the screen.

Going through the rest of the code, you can see that for each entry in the main view, you have to create a dictionary and put it in the `menuList` array. You'll put the dictionary in the `menuList` array so you can use it later when you need to provide the row's content or create a view controller when the user selects the row.

The last thing you do is create the `Destination` object:

```
self.destination =  [[Destination alloc]
                                 initWithName:@"England"];
```

The `Destination` is the model used by the view controllers to get the content needed by the view that is created when the user selects a row. (You will also have to add an `#import "Destination.h"` statement to the `RootViewController.m` file.) I explain the model in detail in Chapter 16.

At this point, you need to add the new instance variables `menuList` and `destination` to `RootViewController.h`.

```
@interface RootViewController : UITableViewController {

  NSMutableArray *menuList;
  Destination    *destination;
}
@property (nonatomic,retain)  Destination    *destination;

@end
```

Because `designation` is a property, you have to add an `@synthesize destination` statement to `RootViewController.h`.

Seeing how cells work

I've been going steadily from macro to micro, so it makes sense that after setting up a model for each row, I get to talk about cells, the individual constituents of each row.

Cell objects are what draw the contents of a row in a table view. The method `tableView:cellForRowAtIndexPath:` is called for each visible row in the table view. It's expected that the method will configure and return a `UITableViewCell` object for each row. The `UITableView` object uses this cell to draw the row.

When providing cells for the table view, you have three general approaches you can take:

✔ Use vanilla (not subclassed) `UITableViewCell` cell objects.

✔ Add subviews to a `UITableViewCell` cell object's content view.

✔ Use cell objects created from a custom subclass of `UITableViewCell`.

The next few sections take a look at these options, one by one.

Using vanilla cell objects

Using the `UITableViewCell` class directly, you can create cell objects with text and an optional image. (If a cell has no image, the text starts near the left edge of the cell.) You also have an area on the right of the cell for accessory views, such as disclosure indicators (the one shaped like a regular chevron), detail disclosure controls (the one that looks like a white chevron in a blue button), and even control objects such as sliders, switches, or custom views. (The layout of a cell is shown in Figure 14-10.) If you like, you can format the font, alignment, and color of the text, as well as have a different format when the row is selected.

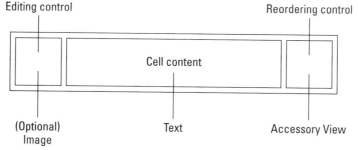

Figure 14-10: The cell architecture.

Editing control — Reordering control — Cell content — (Optional) Image — Text — Accessory View

Adding subviews to a cell's content view

Although you can specify the font, color, size, alignment, and other characteristics of the text in a cell by using the `UITableViewCell` class directly, the formatting is applied to all the text in the cell. To get the variation that I suspect you want between the selection and description text (and, it turns out, the alignment as well), you have to create subviews within the cell.

A cell that a table view uses for displaying a row is, in reality, a view in its own right. `UITableViewCell` inherits from `UIView`, and it has a content view. With content views, you can add one subview (containing, say, the selection text "Weather") formatted the way you want — and a second subview (holding, say, the description text, "Current conditions") formatted an entirely different way. You may remember that you already experienced adding subviews (the button, text field, and labels) in creating the ReturnMeTo application's main view, although you may not have known you were doing that at the time. Well, now it can be told.

Creating a custom subclass UITableViewCell

You can create a custom cell subclass when your content requires it — usually when you need to change the default behavior of the cell.

Creating the cell

As I mention in an earlier section, you're going to use the UITableViewCell class to create the cells for your table views and then add the subviews you need in order to do the formatting you want. The place to create the cell is tableView:cellForRowAtIndexPath:. This method is called for each visible row in the table view, as shown in Listing 14-7. A code stub is already included in the RootViewController.m file, courtesy of the template.

Listing 14-7: Drawing the Text

```
- (UITableViewCell *)tableView:(UITableView * )tableView
        cellForRowAtIndexPath:(NSIndexPath *)indexPath {

  UITableViewCell *cell = [tableView
      dequeueReusableCellWithIdentifier:kCellIdentifier];

  if (cell == nil) {
    cell = [[[UITableViewCell alloc]
          initWithStyle:UITableViewCellStyleDefault
          reuseIdentifier:kCellIdentifier] autorelease];


    cell.accessoryType =
          UITableViewCellAccessoryDisclosureIndicator;

    CGRect subViewFrame = cell.contentView.frame;
    subViewFrame.origin.x += kInset;
    subViewFrame.size.width = kInset+kSelectLabelWidth;

    UILabel *selectLabel = [[UILabel alloc]
                              initWithFrame:subViewFrame];
    selectLabel.textColor = [UIColor blackColor];
    selectLabel.highlightedTextColor = [UIColor
                                        whiteColor];
    selectLabel.font = [UIFont boldSystemFontOfSize:18];
    selectLabel.backgroundColor = [UIColor clearColor];
    [cell.contentView addSubview:selectLabel];

    subViewFrame.origin.x += kInset+kSelectLabelWidth;
    subViewFrame.size.width = kDescriptLabelWidth;
```

```
    UILabel *descriptLabel = [[UILabel alloc]
                            initWithFrame:subViewFrame];
    descriptLabel.textColor = [UIColor grayColor];
    descriptLabel.highlightedTextColor = [UIColor
                                        whiteColor];
    descriptLabel.font = [UIFont systemFontOfSize:14];
    descriptLabel.backgroundColor = [UIColor clearColor];
    [cell.contentView addSubview:descriptLabel];

    int menuOffset = (indexPath.section*kSection1Rows)+
                                    indexPath.row;
    NSDictionary *cellText = [menuList
                            objectAtIndex:menuOffset];

    selectLabel.text = [cellText objectForKey:kSelectKey];
    descriptLabel.text = [cellText
                            objectForKey:kDescriptKey];
    [selectLabel release];
    [descriptLabel release];
    }
return cell;
}
```

Here's the logic behind all that code:

1. **Determine whether there are any cells lying around that you can use.**

   Although a table view can display only a few rows at a time on iPhone's small screen, the table itself can conceivably hold a lot more. A large table would chew up a lot of memory if you were to create cells for every row. Fortunately, table views are designed to *reuse* cells. As a table view's cells scroll off the screen, they're placed in a queue of cells available to be reused.

2. **Create a *cell identifier* that indicates what cell type you're using. Add this to the `Constants.h` file:**

   ```
   #define kCellIdentifier  @"Cell"
   ```

   You can ask the table view for a specific reusable cell object by sending it a `dequeueReusableCellWithIdentifier:` message:

   ```
   UITableViewCell *cell = [tableView
       dequeueReusableCellWithIdentifier:kCellIdentifier];
   ```

   This asks whether any cells of the type you want are available.

3. **If there aren't any cells lying around, you'll have to create a cell, using the cell identifier you just created.**

   ```
   if (cell == nil) {
     cell = [[[UITableViewCell alloc]
         initWithStyle:UITableViewCellStyleDefault
         reuseIdentifier:kCellIdentifier] autorelease];
   ```

   You now have a table-view cell that you can return to the table view.

`UITableViewCellStyleDefault` gives you a simple cell with a text label (black and left-aligned) and an optional image view. There are also several other styles:

- `UITableViewCellStyleValue1` gives you a cell with a left-aligned black text label on the left side of the cell and a smaller blue text and right-aligned label on the right side. (The Settings application uses this style cell.)

- `UITableViewCellStyleValue2` gives you a cell with a right-aligned blue text label on the left side of the cell and a left-aligned black label on the right side of the cell.

- `UITableViewCellStyleSubtitle` gives you a cell with a left-aligned label across the top and a left-aligned label below it in smaller gray text. (The iPod application uses cells in this style.)

4. **Define the accessory type for the cell.**

```
cell.accessoryType =
        UITableViewCellAccessoryDisclosureIndicator;
```

As I mention earlier during my brief tour of a cell, its layout includes a place for an accessory — usually something like a disclosure indicator.

In this case, use `UITableViewCellAccessoryDisclosureIndicator` (the one shaped like a regular chevron). It lets the user know that tapping this entry will result in something (hopefully wonderful) happening — the display of the current weather conditions, for example.

5. **Create the subviews.**

Here I show you just one example. (The other is the same except for the font size and text color.) I get the `contentView` frame and base the subview on it. The inset from the left (`kInset`) and the width of the subview (`kLabelWidth`) are defined in the `Constants.h` file. It looks like this:

```
#define kInset             10
#define kSelectLabelWidth   100
#define kDescriptLabelWidth 160
```

To hold the text, the subview I'm creating is a `UILabelView`, which meets my needs exactly:

```
CGRect subViewFrame = cell.contentView.frame;
subViewFrame.origin.x += kInset;
subViewFrame.size.width = kInset+kSelectLabelWidth;
UILabel *selectLabel = [[UILabel alloc]

            initWithFrame:subViewFrame];
```

You then set the label properties that you are interested in, just as when you created the labels in the ReturnMeTo application. This time, however, you'll do it by manually writing code rather than using Interface Builder. Just set the font color and size — the highlighted font color when an item is selected and the background color of the label (as indicated in the code that follows). Setting the background color to transparent allows me to see the bottom line of the last cell in the group.

```
selectLabel.textColor = [UIColor blackColor];
selectLabel.highlightedTextColor = [UIColor
                                      whiteColor];
selectLabel.font = [UIFont boldSystemFontOfSize:18];
selectLabel.backgroundColor = [UIColor clearColor];
[cell.contentView addSubview:selectLabel];
```

After you have your label, you just set its text to one of the values you get from the dictionary created in awakeFromNib representing this row.

The trouble is that you won't get the absolute row passed to you. You get only the row within a particular section — and you need the absolute row to get the right dictionary from the array. Fortunately, one of the arguments used when this method is called is the indexPath, which contains the section and row information in a single object. To get the row or the section out of an NSIndexPath, you just have to invoke its section method (indexPath.section) or its row method (indexPath.row), each of which returns an int. This neat trick enables you to compute the offset for the row in the array you created in awakeFromNib. This is also why it's so handy to have the number of rows in a section as a constant.

So the first thing you do in the following code is compute that. And then you can use that dictionary to assign the text to the label.

```
int menuOffset = (indexPath.section*kSection1Rows)+
                                      indexPath.row;
NSDictionary *cellText = [menuList
                          objectAtIndex:menuOffset];
selectLabel.text = [cellText objectForKey:kSelectKey];
descriptLabel.text = [cellText
                          objectForKey:kDescriptKey];

menuOffset += indexPath.row;
```

Finally, because I no longer need the labels I created, I release them

```
[selectLabel release];
[descriptLabel release];
```

and return the cell formatted and with the text it needs to display in that row.

```
return cell;
```

If you want it to compile, you also have to comment out the following line of code in awakeFromNib.

```
// self.destination = [[Destination alloc]
                          initWithName:@"England"];
```

Just don't forget to uncomment that line after you have added the Destination class in Chapter 16! If you do compile it now, you can see what I showed you on the left side of Figure 14-1, as shown earlier.

Responding to a selection

When the user taps on a table-view entry, what happens next depends on what you want your table view to do for you.

If this particular application were using the table view to display data (as the Albums view in the iPod application does, for example), you'd show the next level in the hierarchy — such as the list of songs, to stick with the iPod application — or a detail view of an item, such as information about a song.

In this case, you're using the table view as a table of contents, so tapping a table-view entry transfers the user to the view that presents the desired information — the Heathrow Express, for example.

To move from one content view to a new (content) view, first you need to create a new view controller for that view; then you launch it so it creates and installs the view on the screen. But you also have to give the user a way to get back to the main view!

Brass-tacks time: What kind of code-writing gymnastics do you have to do to get all this stuff to happen?

Actually, not much. Table views are usually paired with *navigation bars,* whose job it is to implement the back stuff. And to get a navigation bar, all you have to do is include a *navigation controller* in your application. What's more, if you wisely chose the Navigation-Based Application template at the outset of your iPhoneTravel411 project, a navigation controller was already put in place for you in the appDelegate created by the template. Here's the code that the template quite generously provided you with (the navigation controller is bolded so you can find it easier):

```
@interface iPhoneTravel411AppDelegate : NSObject
                             <UIApplicationDelegate> {

  UIWindow                *window;
  UINavigationController *navigationController;
}

@property (nonatomic, retain) IBOutlet UIWindow *window;
@property (nonatomic, retain) IBOutlet
        UINavigationController *navigationController;
@end
```

This navigation controller is created for you in the MainWindow.xib file (see Figure 14-11), which you can access by double-clicking the MainWindow. xib file in the Groups & Files list in your project. If you take a closer look at

Figure 14-11, you can see that, when the navigation controller is selected, it points to the RootViewController.nib in the View window — which is to say, it's pointing to the RootViewController and its table view. This links together the navigation controller, the root view controller, and the view.

But not only did the Navigation-Based Application template deliver the goods in the iPhoneTravel411AppDelegate.h file and nib file, it also created the code I need in the iPhoneTravel411AppDelegate.m file.

To get the navigation controller view to load in the window, you don't have to do anything. When you chose Navigation-Based Application template, the following code was automatically generated for you:

```
- (BOOL)application:(UIApplication *)
    application didFinishLaunchingWithOptions:
                        (NSDictionary *)launchOptions {

    [window addSubview:navigationController.view];
    [window makeKeyAndVisible];

    return YES;
}
```

When all is said and done, you have a table view with a navigation bar ready to go to work.

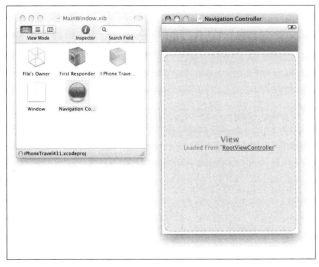

Figure 14-11:
The
navigation
controller.

Navigating the navigation controller

As the preceding section makes clear, to give users the option of returning to a view higher up in the hierarchy (in this case, the main view), table views are paired with navigation bars that enable a user to navigate the hierarchy. Here's what you need to know to make that work:

- The view below the navigation bar presents the current level of data.

- A navigation bar includes a title for the current view.

- If the current view is lower in the hierarchy than the top level, a Back button appears on the left side of the bar; the user can tap it to return to the previous level, as shown in Figure 14-12. The text in the Back button tells the user what the previous level was. In this case, it's the application's main view, so you will see the application's name — iPhoneTravel411.

- A navigation bar may also have an Edit button (on the right side) — used to enter editing mode for the current view — or even custom buttons.

In the case of the iPhoneTravel411 application, you'll create a custom button — an additional Back button. I explain why that is needed shortly.

The navigation bar for each level is managed by a navigation controller — again, as I mention in the preceding section. The navigation controller maintains a stack of view controllers, one for each of the views displayed, starting with the root view controller (hence the name `RootViewController` given to the table-view controller by the template). The root view controller is the very first view controller that the navigation controller pushes onto its stack when a user launches the application; it remains active until the user selects the next view to look at.

Time for a concrete example. When the user taps a row of the table view to get the Heathrow Express information, the root view controller pushes the next view controller onto the stack. The new controller's view (the Heathrow Express information) slides into place and the navigation bar items are updated appropriately. When the user taps the Back button in the navigation bar, the current view controller pops off the stack, the Heathrow View slides off the screen, and the user lands (so to speak) back in the main (table) view.

This is why, in Chapter 8, I decided to place the code to create the keyboard notification in `viewWillAppear`. This is because every time you move back and forth in the hierarchy the `viewWillAppear` message is sent. Although unimportant in ReturnMeTo because there was no hierarchy, in navigation applications if you want something to happen every time a view appears, `viewWillAppear` is the place to do it.

A *stack* is a commonly used data structure that works on the principle of last in, first out. Imagine an "ideal" boarding scenario for an airplane: You would start with the last seat in the last row, and board the plane in back-to-front order until you got to the first seat in the first row — that would be the seat for the last person to board. When you got to your destination you'd deplane (is that really a word?) in the reverse order. That last person on — the person in row one seat one — would be the first person off.

A computer stack is pretty much the same. Adding an object is called a *push* — in this case, when you select Heathrow, the view controller for the Heathrow view is pushed onto the stack. Removing an object is called a *pop* — touching the Back button pops the view controller for the Heathrow view. When you pop an object off the stack, it's always the last one you pushed onto it. The controller that was there before the push is still there, and now becomes the active one — in this case, it's the root view controller.

I mention earlier that I wanted two Back buttons in place. Now you get to find out why. In my design, I wanted to be able to tap a link in the content views to access a Web site. (You can see such a link on the left in Figure 14-12.) When I do that, the iPhoneTravel411 application *replaces* the content of the view, instead of *creating* a new view controller. Tapping the link doesn't change the controller in any way, so the left button doesn't change; you can't use it to get back to a previous view — you only go back to the main view, as the control text tells you. To solve this quandary, I created another button and labeled it "Back" so the user knows he or she can use it to get back to the previous view. I show you how to create such a Back button in Chapter 16.

That being said, Apple's Human Interface Guidelines say, "In addition to displaying web content, a web view provides elements that support navigation through open web pages. Although you can choose to provide webpage navigation functionality, it's best to avoid creating an application that looks and behaves like a mini web browser." In case you want to follow Apple's suggestion here, I show you how to disable links in Chapter 17.

Implementing the selection

At some point, you have to make sure that something actually happens when a user makes a selection. To do that, all you really need to do is implement the `tableview:didSelectRowAtIndexPath:` method to set up a response to a user tap in the main view. This method, too, is already in the `RootViewController.m` file, courtesy of the template. Before I show you that, however, there are some other things you need to put in place that I cover in Chapter 15. Then also in Chapter 15, I explain `tableview:didSele ctRowAtIndexPath:` in detail.

Figure 14-12:
Getting
back.

And Now . . .

You're off to a good start — and you had to use only five methods to create the table and handle user selections. You still have to create the content views and models, but before you do that, I want to show you how to improve the user experience by saving state and allowing the user to set preferences.

Chapter 15

Enhancing the User Experience

"**K**eep the customer satisfied" is my mantra. If that means constantly refining an application design, so be it. In thinking about my iPhone Travel411 design, two things struck me as essential if I really wanted to make this an application that really focuses on the user. The first is part of the Human Interface Guidelines, so it's not really something I can claim credit for; the second is something that flowed straight out of the nature of my design.

In this chapter, I show how I incorporated elements into my design that directly addressed issues relating to an enhanced user experience.

Saving and Restoring State

In computer science terms, a *state* is a unique configuration of information in a program or machine. For our purposes, if the user leaves the application because he or she got a phone call or SMS message, or even decided to play a game, you want the user, when he or she resumes the application, to be able to start exactly where he or she left off.

You may be wondering to yourself, "Why do I have to save state? Didn't you say way back in Chapter 6 that under iOS 4, when the user taps the Home button on a device, the application is suspended and when you 'launch' it again, it starts right back up where it left off."

Yes, I did say that, but there are two situations where that won't happen:

✔ The user is running a device that doesn't support multitasking. (There are still a lot of iPhone and iPod Touches that are not running iOS 4 out there.)

✔ The device does support multitasking, but your app has been purged from memory.

If either of these situations crop up, that means you have to save any unsaved data — as well as the current state of your application — if you want to restore the application to its previous state the next time the user launches it. Now, in situations like this one, you have to use common sense to decide what *state* really means. Generally, you wouldn't need to restore the application to where the user last stopped in a scrollable list, for example. For purposes of explanation, I chose to save the last category view that the user selected in the main table view, which corresponds to a row in a section in the table view. You, the reader, might also consider saving that last view that was selected in that category.

So where do you do that?

In those devices that don't support multitasking, when the user taps the Home button, the iPhone OS terminates your application and returns to the Home screen. The `applicationWillTerminate:` message is sent, and your application is terminated — no ifs, ands, or buts. That's the place for devices that don't support multitasking.

In devices that do support multitasking, the `applicationWillTerminate:` message is not sent. Instead, when your app is moved into background, the `applicationDidEnterBackground:` message is sent, and you have to save any changes in this method in case your application is later purged.

Saving state information

Here's the sequence of events that go into saving the state:

1. **As is shown in bold in Listing 15-1, add a new instance variable `lastView` and declare the `@property` in the `iPhoneTravel411AppDelegate.h` file. Also add the `saveState` declaration. (You'll implement that shortly.)**

   I explain properties in Chapter 8.

   As you can see, `lastView` is a mutable array. You'll save the section as the first element in the array and the row as the second element. As I mention in Chapter 14, because it's mutable, it'll be easier to update when the user selects a new row in a section.

2. **As is shown in Listing 15-2, add the `@synthesize` statement to the `iPhoneTravel411AppDelegate.m` file, to tell the compiler to create the accessors for you.**

   This is shown in Listing 15-2. (You guessed it — new stuff is bold.)

3. **Define the filename you'll use when saving the state information in the `Constants.h` file.**

   ```
   #define kState  @"LastState"
   ```

You also have to add the #import "Constants.h" statement to the iPhoneTravel411AppDelegate.h file.

4. **Save the section and row that the user last tapped in the iPhone Travel411AppDelegate's lastView instance variable by adding the following code to the beginning of the tableview:didSelectRowAt IndexPath: method in the RootViewController.m file, as shown in Listing 15-3 (and in context in Listing 15-13).**

The tableview:didSelectRowAtIndexPath: method is called when the user taps a row in a section. As you recall from Chapter 14, the section and row information are in the indexPath argument of the tab leview:didSelectRowAtIndexPath: method. All you have to do to save that information is to save the indexPath.section as the first array entry, and the indexPath.row as the second. (The reason I do it this way will become obvious when I show you how to write this to a file.)

You also have to add #import "iPhoneTravel411AppDelegate.h".

5. **When the user goes back to the main view, save that main view location in the viewWillAppear: method. You need to add this method to the RootViewController.m file, as shown in Listing 15-4. (It's already there; all you have to do is uncomment it out.)**

The last step is to deal with the case when the user moves back to the main view and then quits the application. To indicate that the user is at the main view, I use –1 to represent the section and –1 to represent the row. I use minus ones in this case because, as you recall, the first section and row in a table are both 0, which requires me to represent the table (main) view itself in this (clever) way.

6. **Save the section and row in the saveState method by adding the code in Listing 15-5.**

In saveState, I'm saving the lastView instance variable (which contains the last section and row the user tapped) to the file kState, which is the constant I defined in Step 3 to represent the filename LastState.

As you can see, reading or writing to the file system on the iPhone is pretty simple: You tell the system which directory to put the file in, specify the file's name, and then pass that information to the write ToFile method. Let me take you through what I just did in Step 6:

- *I got the path to the Documents directory.*

```
NSArray *paths = NSSearchPathForDirectoriesInDomains
        (NSDocumentDirectory, NSUserDomainMask, YES);
NSString *documentsDirectory =[paths objectAtIndex:0];
```

On the iPhone, you really don't have much choice about where the file goes. Although there's a /tmp directory, I'm going to place this file in the Documents directory — because (as I explain in Chapter 2), this is part of my application's sandbox, so it's the natural home for all the app's files.

NSSearchPathForDirectoriesInDomains: returns an array of directories; because I'm only interested in the Documents directory, I use the constant NSDocumentDirectory — and because I'm restricted to my home directory, /sandbox, the constant NSUserDomainMask limits the search to that *domain*. There will be only one directory in the domain, so the one I want will be the first one returned.

- *I created the complete path by appending the path filename to the directory.*

```
NSString *filePath = [documentsDirectory
              stringByAppendingPathComponent:fileName];
```

stringByAppendingPathComponent; precedes the filename with a path separator (/) if necessary.

Unfortunately, this doesn't work if you're trying to create a string representation of a URL.

- *I wrote the data to the file.*

```
[lastView writeToFile:filePath atomically:YES];
```

writeToFile: is an NSData method and does what it implies. I'm actually telling the array here to write itself to a file, which is why I decided to save the location in this way in the first place. There are a number of other classes that implement that method, including NSData, NSDate, NSNumber, NSString, and NSDictionary. You can also add this behavior to your own objects, and they could save themselves — but I won't get into that here. The atomically parameter first writes the data to an auxiliary file, and when that's successful, it's renamed to the path you've specified. This guarantees that the file won't be corrupted even if the system crashed during the write operation.

7. **Send the saveState message in both the applicationDidEnter Background: and applicationWillTerminate: methods.**

The result is shown in Listing 15-6.

Listing 15-1: Adding the Instance Variable to the Interface

```
@interface iPhoneTravel411AppDelegate : NSObject
                                <UIApplicationDelegate> {

    UIWindow                 *window;
    UINavigationController   *navigationController;
    NSMutableArray           *lastView;

}

@property (nonatomic, retain) IBOutlet UIWindow *window;
@property (nonatomic, retain) IBOutlet
```

```
                    UINavigationController *navigationController;
@property (nonatomic, retain) NSMutableArray *lastView;

- (void) saveState;

@end
```

Listing 15-2:   Adding the @synthesize to the Implementation

```
#import "iPhoneTravel411AppDelegate.h"
#import "RootViewController.h"
#import "Constants.h"

@implementation iPhoneTravel411AppDelegate

@synthesize window;
@synthesize navigationController;
@synthesize lastView;
```

Listing 15-3:   Saving indexPath

```
iPhoneTravel411AppDelegate *appDelegate =
          (iPhoneTravel411AppDelegate *)[[UIApplication
          sharedApplication] delegate];
[appDelegate.lastView replaceObjectAtIndex:0
          withObject:[NSNumber
          numberWithInteger:indexPath.section]];
[appDelegate.lastView replaceObjectAtIndex:1
          withObject:[NSNumber
          numberWithInteger:indexPath.row]];
```

Listing 15-4:   Adding viewWillAppear:

```
- (void)viewWillAppear:(BOOL)animated {

  iPhoneTravel411AppDelegate *appDelegate =
          (iPhoneTravel411AppDelegate *)
          [[UIApplication sharedApplication] delegate];

  [appDelegate.lastView replaceObjectAtIndex:0
           withObject:[NSNumber numberWithInteger:-1]];
  [appDelegate.lastView replaceObjectAtIndex:1
           withObject:[NSNumber numberWithInteger:-1]];
}
```

Listing 15-5: Adding saveState

```
- (void) saveState {

    NSArray *paths = NSSearchPathForDirectoriesInDomains
                (NSDocumentDirectory, NSUserDomainMask, YES);
    NSString *documentsDirectory = [paths objectAtIndex:0];
    NSString *filePath = [documentsDirectory
                    stringByAppendingPathComponent:kState];
    [lastView writeToFile:filePath atomically:YES];
}
```

Listing 15-6: Updating applicationDidEnterBackground: and applicationWillTerminate:

```
- (void) applicationDidEnterBackground:(UIApplication *)
                                            application {

    [self saveState];
}

- (void) applicationWillTerminate:(UIApplication *)
                                            application {

    [self saveState];
}
```

Restoring the state

Now that I've saved the state, I need to restore it when the application is launched. I use my old friend `application;didFinishLaunchingWithOptions:` to carry out that task (as shown in Listing 15-7). `application;didFinishLaunchingWithOptions:` is a method you can find in the `iPhoneTravel411AppDelegate.m` file. The code you need to add is in bold.

Listing 15-7: Adding to application:didFinishLaunchingWithOptions:

```
- (BOOL) application:(UIApplication *)
        application didFinishLaunchingWithOptions:
                            (NSDictionary *)launchOptions {

    NSArray *paths = NSSearchPathForDirectoriesInDomains
            (NSDocumentDirectory, NSUserDomainMask, YES);
    NSString *documentsDirectory = [paths objectAtIndex:0];
    NSString *filePath = [documentsDirectory
                    stringByAppendingPathComponent:kState];
    lastView =[[NSMutableArray alloc]  initWithContentsOfFil
            e:filePath];
    if (lastView == nil) {
```

```
lastView = [[NSMutableArray arrayWithObjects:
                [NSNumber numberWithInteger:-1],
                [NSNumber numberWithInteger:-1],
                nil] retain];
}
[window addSubview:[navigationController view]];
[window makeKeyAndVisible];
}
```

Reading is the mirror image of writing. I create the complete path, including the filename, just as I did when I saved the file. This time I send the `initWith ContentsOfFile:` message instead of `writeToFile:`, which allocates the `lastView` array and initializes it with the file. If the result is `nil`, there's no file, meaning that this is the first time the application is being used. In that case, I create the array with the value of section and row set to −1 and −1. (As I mention earlier, in Step 5, I use −1 −1 to indicate the main view because 0 0 is actually the first row in the first section.)

TIP

`initWithContentsOfFile:` is an `NSData` method similar to `writeTo File:`. The classes that implement `writeToFile:` and those that implement `initWithContentsOfFile:` are the same.

Fortunately, restoring the current state is actually straightforward, given the program architecture. The `RootViewController`'s `viewDidLoad` method is called at application launch — after the first view is in place but not yet visible. At that point, you're getting ready to display the (table) view. But instead of just doing that, you see whether the saved view was something other than the table view, and if it was, you take advantage of the same mechanisms that are used when the user taps a row in the table view. You invoke the `didSelectRowAtIn dexPath:` method, which already knows how to display a particular view represented by the `indexPath`, that is, section and row. This is shown in Listing 15-8.

`viewDidLoad` is already in the `RootViewController.m` file. All you have to do is uncomment it out.

Listing 15-8:   Specifying the View to be Displayed at Launch

```
- (void)viewDidLoad {

    iPhoneTravel411AppDelegate *appDelegate =
                (iPhoneTravel411AppDelegate *)
                [[UIApplication sharedApplication] delegate];

    if ([[((NSNumber*) [appDelegate.lastView
                        objectAtIndex:0]) intValue] != -1) {
        NSIndexPath* indexPath = [NSIndexPath indexPathForRow:
            [[appDelegate.lastView objectAtIndex:1]intValue]
            inSection:
```

(continued)

Listing 15-8 *(continued)*

```
        [[appDelegate.lastView objectAtIndex:0] intValue]];
    [self tableView:((UITableView*) self.tableView)
                        didSelectRowAtIndexPath:indexPath];
    }
}
```

Here's what you're up to in Listing 15-8:

1. **Check to see whether the last view was the table view.**

   ```
   if ([[((NSNumber*) [appDelegate.lastView
               objectAtIndex:0]) intValue] != -1) {
   ```

2. **If the last view wasn't the table view, create the index path using the last section and row information that was loaded into the `lastView` instance variable by `application;didFinishLaunchingWithOptions:`.**

   ```
   NSIndexPath* indexPath = [NSIndexPath indexPathForRow:
           [[appDelegate.lastView objectAtIndex:1]
           intValue] inSection:
       [[appDelegate.lastView objectAtIndex:0] intValue]];
   ```

3. **Send the `tableview:didSelectRowAtIndexPath:` message to display the right view.**

   ```
   [self tableView:((UITableView*) self.tableView)
                       didSelectRowAtIndexPath:indexPath];
   ```

The reason I created an index path was to be able to take advantage of the `didSelectRowAtIndexPath:` method to replay the last user tap in the main view.

Respecting User Preferences

The left side of Figure 15-1 shows you the Settings screen for my Mobile Travel411 application. There you can see that I've provisioned my app with four preferences. The right side of Figure 15-1 shows the Settings screen for iPhoneTravel411. In this chapter, I show you how to implement the most critical of these preferences — the Use Stored Data option. There's also another preference, Monitor Distance, that I haven't discussed much and was not included in the original MobileTravel411 design. Being able to easily and efficiently track a user's location and use that information to enhance the user experience is an important feature in iOS 4. So like any good developer, I've enhanced my app. (As for how one goes about adding location-based features, I show you how to do that in Chapter 18.)

Use Stored Data tells the application to use the last version of the data that it accessed instead of going out on the Internet for the latest information.

Although this does violate my I Want The Most Up-To-Date Information guideline, it can save the user from excessive roaming charges, depending on his or her cell provider's data plan.

Figure 15-1:
The
required
preferences.

 No doubt it's way cool to put user preferences in Settings. Some programmers abuse this trick, though; they make you go into Settings, when it's just as easy to give the user a preference-setting capability within the program itself. You should put something in Settings only if the user changes it infrequently. In this case, stored data doesn't change often; Both Use Stored Data mode and Monitor Distance definitely belongs in Settings.

Just a heads up: In this part of the chapter, I show you how to put a toggle switch in Settings that lets you specify whether to use only stored data — and then I show you how to retrieve the setting — but you have to jump to Chapter 18 to find out how to use Monitor Distance. In this and in the next chapter, I show you how to actually use the toggle switch setting in your code.

The Settings application uses a property list, called `Root.plist`, found in the Settings bundle inside your application. The Settings application takes what you put in the property list and builds a Settings section for your application in its list of application settings as well as the views that display and enable the user to change those settings. The next sections spell out how to put that Settings section to work for you.

Adding a Settings bundle to your project

For openers, you have to add a Settings bundle to your application. Here are the moves:

1. **In the Groups & Files list (at left in the Xcode Project window), select the iPhoneTravel411 folder and then chose File⇨New File from the main menu, or press ⌘+N.**

   The New File dialog appears.

2. **Choose Resource under the iPhone OS heading in the left pane, and then select the Settings Bundle icon, as shown in Figure 15-2.**

3. **Click the Next button.**

4. **Choose the default name of `Settings.bundle` and then press Return (Enter) or click Finish.**

   You should now see a new item called `Settings.bundle` in the iPhoneTravel411 folder, in the Groups & Files list.

5. **Click the triangle to expand the `Settings.bundle` subfolder.**

   You see the `Root.plist` file as well as an `en.lproj` folder — the latter is used for dealing with localization issues, as I discuss in Chapter 14.

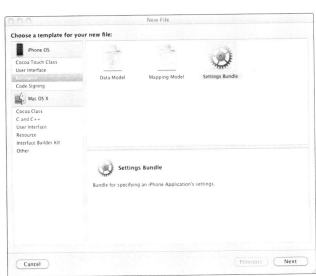

Figure 15-2:
Creating the application bundle.

Setting up the property list

Property lists are widely used in iPhone applications because they provide an easy way to create structured data using named values for a number of object types.

In the MobileTravel411 application, I use property lists extensively as a way to *parameterize* view controllers and models. (I de-buzz this word and provide details in the next chapter.)

Property lists all have a single root node — a Dictionary, which means it stores items using a key-value pair, just as an NSDictionary does: All dictionary entries must have both a key and a value. In this dictionary, there are two keys:

- ✔ StringsTable
- ✔ PreferenceSpecifiers

The value for the first entry is a string — the name of a strings table used for localization, which I don't get into here. The second entry is an array of dictionaries — one dictionary for each preference. (You probably need some time to wrap your head around that one; it'll become clearer as I take you through it.)

PreferenceSpecifiers is where you put a toggle switch so the user can choose to use (or not use, because it's a toggle) only stored data — I refer to that choice later as *stored data mode*. Here's how it's done:

1. **In the Groups & Files list of the Project window, select the disclosure triangle next to the `Settings.bundle` file to reveal the `Root.plist` file, and then double-click the `Root.plist` file to open it in a separate window, as shown in Figure 15-3.**

   Okay, you don't *really* have to do this, but I find it easier to work with this file when it's sitting in its own window.

2. **In the `Root.plist` window you just opened, expand the disclosure triangles next to all the nodes by clicking all those triangles, as shown in Figure 15-3.**

   You can also expand everything by holding down the Option key when clicking a closed disclosure triangle, like the one next to PreferenceSpecifiers.

3. **Under the `PreferenceSpecifiers` heading in the `Root.plist` window, move to Item 0.**

   PreferenceSpecifiers is an array designed to hold a set of dictionary nodes, each of which represents a single preference. For each item listed in the array, the first row under it has a key of Type; every property list node in the `PreferenceSpecifiers` array must have an entry with this key, which identifies what kind of entry this is. The `Type` value for the current Item 0 — `PSGroupSpecifier` — is used to indicate that a new group should be started. The value for this key actually acts like a section heading for a table view (like you created in Chapter 14). Double-click the value next to `Title` and delete the default `Group`, as I have in Figure 15-4 (or you can put in `IPhoneTravel411 Preferences`, or be creative if you like).

4. **Seeing that Item 2 is already defined as a toggle switch, you can just modify it by changing the `Title` value from `Enabled` to `Use stored data` and the key from `enabled_preference` to `useStoredData Preference`.**

   This is the key you'll use in your application to access the preference.

5. **Continue your modifications to Item 2 by deselecting the `Boolean` check box next to `DefaultValue`.**

   I want the Use Stored Data preference initially to be set to Off because I expect most people will still want to go out on the Internet for the latest information, despite the high roaming charges involved.

   When you're done, the `Root.plist` window should look like Figure 15-4.

6. **Collapse the disclosure triangles next to items 1 and 3 (shown in Figure 15-5) and then select those items one by one and delete them.**

   The item numbers do change as you delete them, so be careful. That's why you need to leave the preference item you care about open, so you can see that you shouldn't delete it. Fortunately, Undo is supported here; if you make a mistake, press ⌘+Z to undo the delete.

7. **Save the property list file by pressing ⌘+S.**

As I mention at the beginning of this section. I want you to add another preference — Monitor Distance. To do that, collapse the disclosure triangle next to Item 1, select Item 1, and then copy and paste it into the file. (Press ⌘+C and then ⌘+V.) (Item 1 of course used to be Item 2, before you did that deleting business in Step 6.) You'll see that you have added a new `PreferenceSpecifier` — a brand spanking new Item 2. Expand the disclosure triangle next to Item 2 and replace `Use Stored Data` in the `Title` field with `Monitor Distance` and `useStoredDataPreference` in the Key field with `monitorLocationPreference`. Also, this time select the `Boolean` check box in the DefaultValue field. When you're done, your `root.plist` should look like Figure 15-6.

Be sure to save the property list file by pressing ⌘+S.

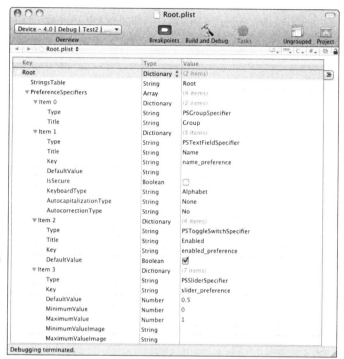

Figure 15-3:
The default
Root.plist
file
preferences.

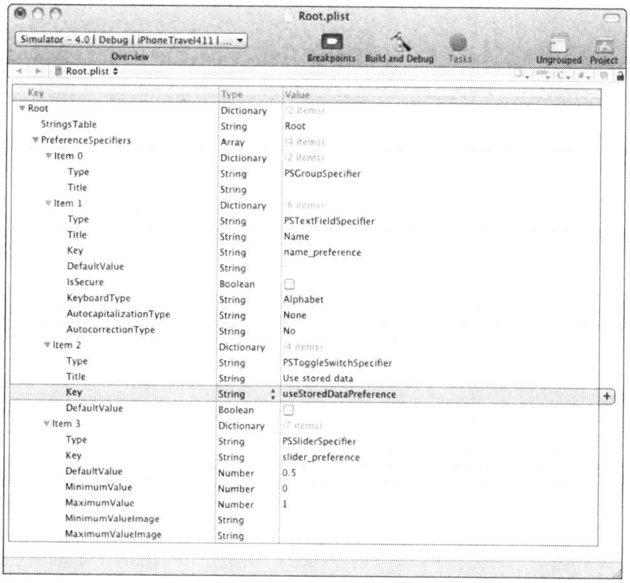

Figure 15-4:
Preferences
for IPhone
Travel411.

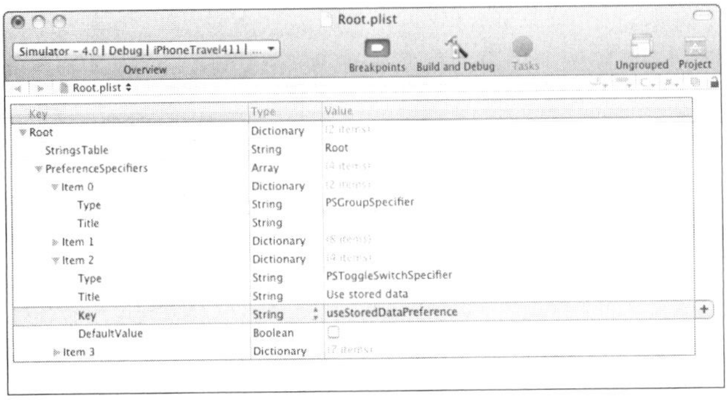

Figure 15-5:
Delete these
items.

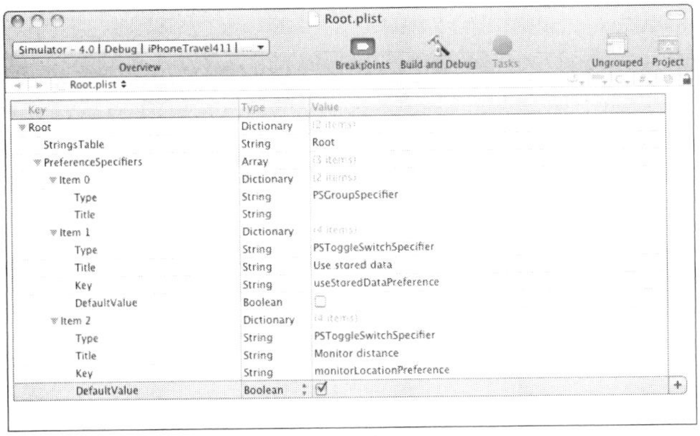

Figure 15-6:
Adding a
new
preference.

Reading Settings in the Application

After you set it up so your users can let their preferences be known in Settings, you need to read those preferences back into the application. You do that in the `iPhoneTravel411AppDelegate`'s `application:didFinishLaunchingWithOptions:` method. But first, a little housekeeping.

1. **First add the two new instance variables `useStoredData` and `monitorLocationchanges` and then declare them as properties in the `iPhoneTravel411AppDelegate.h` file.**

   This is shown in Listing 15-9. (Again, the new stuff is bold.)

   Notice that the `@property` declaration is a little different than what you have been using so far. Up to now, all your properties have been declared `(nonatomic, retain)` — as is explained in Chapter 7. What's this `readonly` stuff? Because both `useStoredData` and `monitorLocationChanges` are not objects (they're Boolean values), `retain` is not applicable. In addition, you'll enable it to be read-only. If you wanted it to be updatable, you could make it `readwrite`.

2. **Add the @synthesize statements to the iPhoneTravel411AppDelegate.m file, to tell the compiler to create the accessors for you.**

This is shown in Listing 15-10. (You guessed it — new is bold.)

Listing 15-9:   Adding the Instance Variable to the Interface

```
@interface iPhoneTravel411AppDelegate : NSObject
                                  <UIApplicationDelegate> {

    UIWindow                 *window;
    UINavigationController   *navigationController;
    NSMutableArray           *lastView;
    BOOL                     useStoredData;
    BOOL                     monitorLocationChanges;
}

@property (nonatomic, retain) IBOutlet UIWindow *window;
@property (nonatomic, retain) IBOutlet
          UINavigationController *navigationController;
@property (nonatomic, retain) NSMutableArray *lastView;
@property (nonatomic, readonly) BOOL useStoredData;
@property (nonatomic, readonly) BOOL
                                  monitorLocationChanges;

@end
```

Listing 15-10:   Adding the @synthesize to the Implementation

```
#import "iPhoneTravel411AppDelegate.h"
#import "RootViewController.h"
#import "Constants.h"

@implementation iPhoneTravel411AppDelegate

@synthesize window;
@synthesize navigationController;
@synthesize lastView;
@synthesize useStoredData;
@synthesize monitorLocationChanges;
```

With your housekeeping done, it's time to add the necessary code. But first, I want to explain something about defaults in a world of multitasking.

As I explain in Chapter 6, when the user is (temporarily) done with your application, the application does not terminate. Instead it goes into background and becomes inactive. But while your application is sitting there in the background, life goes on, and the user may have done something that impacts your application, like change the settings. (There are a few other things that you need to pay attention to, and I explain them in Chapter 18.)

Fortunately, iOS 4 provides a way for you to be informed of what has happened while your application dreams of electric sheep. While your app is suspended, the user could be doing all sorts of things — like getting on a plane to London and changing his or her preference to use stored data rather than the Internet. Although I'm going to concentrate on changes in user preferences, here's a list of some of the other things that could potentially impact your app:

- An accessory is connected or disconnected
- The device orientation changes
- There is a significant time change
- The battery level or battery state changes
- The proximity state changes
- The status of protected files changes
- An external display is connected or disconnected
- The screen mode of a display changes
- Preferences that your application exposes through the Settings application are changed
- The current language or locale settings changes

That's a lot to keep track of, but iOS 4 is very helpful in keeping things straight for you. Instead of saving all those events and pummeling your application senseless with everything the user has done for the last six weeks, it coalesces events and delivers a single event (of each relevant type) that nets out all the changes since your app was suspended.

The way you find out if anything has changed is through our old friend the Notification System that I explain in Chapter 8. These net changes are sent to you in the `NSUserDefaultsDidChangeNotification` notification. For example, in iPhoneTravel411 the user may have turned on Use Shared Data. If you didn't respond to that when your application became active again, you could potentially rack up a rather large roaming bill for the user.

If you do receive the `NSUserDefaultsDidChangeNotification` notification, the appropriate response would be to reload the settings data and have your app behave appropriately.

To get that notification, though, you need to register for the `NSUserDefaultsDidChangeNotification` in the `application:didFinishLaunchingWithOptions:` method. Right after that, you'll want to read in the user preferences and set your new instance variables accordingly. Listing 15-11 shows you how it's done.

Listing 15-11: Adding to application:didFinishLaunchingWithOptions

```
- (BOOL)application:(UIApplication *)
    application didFinishLaunchingWithOptions:
                             (NSDictionary *)launchOptions {

  [[NSNotificationCenter defaultCenter] addObserver:self
    selector:@selector(setDefaults:)
    name:NSUserDefaultsDidChangeNotification object:nil];

  if (![[NSUserDefaults standardUserDefaults]
       objectForKey:kUseStoredDataPreference]) {
    useStoredData = NO;
    monitorLocationChanges = YES;
}
  else
    [self setDefaults:nil];

  NSArray *paths = NSSearchPathForDirectoriesInDomains
          (NSDocumentDirectory, NSUserDomainMask, YES);
  NSString *documentsDirectory = [paths objectAtIndex:0];
  NSString *filePath = [documentsDirectory
                        stringByAppendingPathComponent:kSt
          ate];

  lastView =[[NSMutableArray alloc]  initWithContentsOfFil
          e:filePath];
  if (lastView == nil) {
    lastView = [[NSMutableArray arrayWithObjects:
                [NSNumber numberWithInteger:-1],
                [NSNumber numberWithInteger:-1],
                nil] retain];
  }
  [window addSubview:navigationController.view];
  [window makeKeyAndVisible];

  return YES;
}
```

Here's what you want all that code to do for you:

1. **Register for the `NSUserDefaultsDidChangeNotification`. Inform the Notification Center to send you the `setDefaults:` message in the event the user changes a preference.**

   This is essentially the same thing you did in Chapter 8 when you registered for the UIKeyboardWillShowNotification.

   ```
   [[NSNotificationCenter defaultCenter] addObserver:self
     selector:@selector(setDefaults:)
     name:NSUserDefaultsDidChangeNotification object:nil];
   ```

2. **Check to see whether the settings have been moved into `NSUserDefaults`.**

```
if (![[NSUserDefaults standardUserDefaults]
           objectForKey:kUseStoredDataPreference]){
```

I explain `NSUserDefaults` back in Chapter 9. The Settings application moves the user's preferences from Settings into `NSUserDefaults` only *after* the user visits the setting for the first time. (I want to point out that if the user visits the setting and doesn't change anything, you won't get a change notification either.)

3. **If the settings haven't been moved into `NSUserDefaults` yet, then use the defaults of `NO` and `YES` (which corresponds to the default you used for the initial preference value).**

```
useStoredData = NO;
monitorLocationChanges = YES;
```

4. **If the settings *have* been moved into `NSUserDefaults`, send yourself the message**

```
[self setDefaults:nil];
```

This is the same message you signed up to have sent when you registered for the `NSUserDefaultsDidChangeNotification`. In this case, you use `nil` as the argument because it isn't coming from the Notification Center with a notification. Add the `setDefaults:` message shown in Listing 15-12.

You also have to add the following to `Constants.h`:

```
#define kUseStoredDataPreference
            @"useStoredDataPreference"
#define kMonitorLocationPreference
            @"monitorLocationPreference"
```

Listing 15-12:   setDefaults:

```
- (void)setDefaults:(NSNotification*)notification {

  useStoredData = [[NSUserDefaults standardUserDefaults]
                   boolForKey:kUseStoredDataPreference];
  monitorLocationChanges =
         [[NSUserDefaults standardUserDefaults]
               boolForKey:kMonitorLocationPreference];
}
```

All you do here is read in the defaults from the file, just like what you did back in Chapter 9 when you read in the saved phone number for the ReturnMeTo app. In this case, you don't need to use the notification argument.

Using Preferences in Your Application

I say back in Chapter 14 that before I could explain the `tableview:didSele
ctRowAtIndexPath:` method — the method that makes something happen
when a user selects a row in the table view — you need to have some other
things in place. And while there are other places in your app where you'll
need to know if you are in stored data mode, the `tableview:didSelectRo
wAtIndexPath:` method really needs to know if you are in that mode. Now
that you have implemented that preference, take a look at the entire method
in Listing 15-13.

Listing 15-13:   tableview:didSelectRowAtIndexPath:

```
- (void)tableView:(UITableView *)tableView
        didSelectRowAtIndexPath:(NSIndexPath *)indexPath {

  [tableView deselectRowAtIndexPath:indexPath
          animated:YES];

  int menuOffset = (indexPath.section*kSection1Rows)+
          indexPath.row;
  iPhoneTravel411AppDelegate *appDelegate =
          (iPhoneTravel411AppDelegate *)
            [[UIApplication sharedApplication] delegate];
  [appDelegate.lastView replaceObjectAtIndex:0 withObject:
          [NSNumber numberWithInteger:indexPath.section]];
  [appDelegate.lastView replaceObjectAtIndex:1 withObject:
            [NSNumber numberWithInteger:indexPath.row]];

  NSLog (@" %@", [[menuList objectAtIndex:menuOffset]
          objectForKey:kSelectKey]);

  UIViewController *targetController =
          [[menuList objectAtIndex:menuOffset]
          objectForKey:kControllerKey];

  if (targetController == nil) {
    iPhoneTravel411AppDelegate *appDelegate =
          (iPhoneTravel411AppDelegate *) [[UIApplication
          sharedApplication] delegate];
    BOOL realtime = !appDelegate.useStoredData;

    switch (menuOffset) {
      case 0:
        targetController = [[CityController alloc] initWit
          hDestination:destination];
        break;
      case 1:
```

(continued)

Listing 15-13 *(continued)*

```
        if (realtime) targetController = [[MapController
            alloc] initWithDestination:destination];
        else [self displayOfflineAlert:[[menuList
            objectAtIndex:menuOffset]
            objectForKey:kSelectKey]];
        break;
    case 2:
        targetController = [[CurrencyController alloc]
            initWithDestination:destination];
        break;
    case 3:
        if (realtime) targetController =
                    [[WeatherController alloc]
                    initWithDestination:destination];
        else [self displayOfflineAlert:[[menuList
            objectAtIndex:menuOffset]
            objectForKey:kSelectKey]];
        break;
    case 4:
        targetController = [[AirportController alloc]
            initWithDestination:destination airportID:1];
        break;
    case 5:
        targetController = [[AirportController alloc]
            initWithDestination:destination airportID:2];
        break;
    case 6:
        targetController = [[AirportController alloc]
            initWithDestination:destination airportID:2];
        break;
    }
    if (targetController) {
      [[menuList  objectAtIndex:(indexPath.row +
          (indexPath.section*kSection1Rows))]
      setObject:targetController forKey:kControllerKey];
      [targetController release];
    }
  }
  if (targetController) [[self navigationController]
  pushViewController:targetController animated:YES];
}
```

Here's what happens when a user makes a selection in the main view:

1. **You deselect the row the user selected.**

```
[tableView deselectRowAtIndexPath:indexPath
                                animated:YES];
```

It stands to reason that if you want your app to move on to a new view, you have to deselect the row where you currently are.

2. You compute the offset (based on section and row) into the menu array.

```
int menuOffset =
        (indexPath.section*kSection1Rows)+ indexPath.row;
```

You need to figure out where you want your app to land, right?

3. You save the section and row that the user last tapped.

I covered that in Step 4 in the "Saving state information" section, earlier in this chapter.

4. You check to see whether the controller for that particular view has already been created.

```
UIViewController *targetController =
                [menuList objectAtIndex:menuOffset]
                objectForKey:kControllerKey];
if (targetController == nil) {
```

5. If no controller exists, you create and initialize a new controller if needed.

I explain the mechanics of creating and initializing a new controller in Chapter 16. As you can see, you're going to use another `switch` statement to get to the right controller:

```
switch (menuOffset) {
```

For many of the selections, you'll always create a new controller. For example:

```
case 4:
  targetController = [[AirportController alloc]
        initWithDestination:destination airportID:1];
  break;
```

But for some selections, like Weather, you have decided that if you're not online, you can't deliver the quality of the information a user needs. (Saved current weather conditions is an oxymoron.) For other selections — Map, for example — a network connection is required. (Right now, no caching is available.) In that case, you send an alert to the user (see Listing 15-14) informing him or her that the selection is unavailable. (You also need to add the declaration to `RootViewController.h`.)

```
if (realtime) targetController =
                [[MapController alloc]
                initWithDestination:destination];
else [self displayOfflineAlert:
        [[menuList objectAtIndex:menuOffset]
        objectForKey:kSelectKey]];
```

6. **If you created a new view controller, you save a reference to the newly created controller in the dictionary for that row.**

```
if (targetController) {
    [[menuList objectAtIndex
    (indexPath.row + (indexPath.section*kSection1Rows))]
    setObject:targetController forKey:kControllerKey];
    [targetController release];

}
```

7. **If you created a new view controller, you push the controller onto the stack and let the navigation controller do the rest.**

```
if (targetController) [[self navigationController]
    pushViewController:targetController animated:YES];
```

Listing 15-14:   Displaying an Alert

```
- (void) displayOfflineAlert: (NSString*) selection {

    UIAlertView *alert = [[UIAlertView alloc]
            initWithTitle:selection
            message:@"is not available offline"
            delegate:self cancelButtonTitle: @"Thanks"
            otherButtonTitles:nil];
    [alert show];
    [alert release];
}
```

Of course, if you add all this code and try to compile your application this very minute, you're going to get a slew of errors. You haven't yet defined the view controllers you'll need. So, if you do want to compile the application and see the results, be sure to comment out everything that refers to classes you haven't created yet. Just be sure to go back and uncomment them all out after you *do* create the controllers in the next chapter!

This App Is Almost Done

In the next two chapters, I show you how to implement the content view controllers, views, and models. Then you'll be ready to march off and create your own applications.

Chapter 16

Creating Controllers and Their Models

*G*etting the infrastructure in place for a new iPhone application is certainly a crucial part of the development process, but in the grand scheme of things, it's only the spadework that prepares the way for the really cool stuff. After all is said and done, you still need to add the content that the users will see or interact with. Content is, after all, the reason they bought this application.

Chapter 15 was all about infrastructure — and a very interesting chapter it was — but this chapter moves on from there to content views and how to implement them.

This is actually less difficult than it sounds. The real key to creating an extensible application — one where you can easily add new views — is the program architecture. When you have a table view in place, you have the structure necessary for navigation in your application. When you already have that navigation architecture set up — along with the MVC pattern I've been touting all along — creating the views, view controllers, and the model turns out to be somewhat pedestrian. You do more or less the same thing over again for each new content view. (Oh well, boring is sometimes good.)

If you've been dutifully following along in this book chapter by chapter, you probably already know the basics of creating a view and its view controller. If so, it's no big surprise that what I do here is almost identical to creating the view controller for my other sample application (the ReturnMeTo app) in Chapters 5, 7, 8, 9 and 11 — although this time I get to do it in a single chapter!

To kick off the process, you first need to decide what you want to see in each view whenever the user selects a particular row in the main view. You also need to decide where that information is going to be located. You have a number of options here. It can be a program resource (kind of like a local file, which I get to later), a Web page, or data located on a server on the Internet. To ease your mind, I show you how to work with all three. And, oh yes, as you started to do in Chapter 15, you also have to make some decisions about what you want to do if the user is offline.

My running examples in this chapter are going to involve implementing the Currency, Weather, and Heathrow selections for the iPhoneTravel411 application, along with an introduction to the Map selection that may surprise you. (I cover the Map selection in detail in Chapter 17.) To see the full listings of each of those selections, you have to go to my Web site — `www.neal goldstein.com`. At this stage of the game, the listings have gotten so big that it would be a bear for you to flip back and forth between them in printed-text form. Having them in electronic form also means that, if you download them, you can create an Xcode project and work along and experiment with changing parameters or even logic just to see what happens — very helpful strategies if you want to find out what's *really* going on.

You'll work with the project in the folder named iPhoneTravel411 Chapter 16, which will have the code for the finished application through Chapter 16. The iPhoneTravel411 Chapter 17 folder will have the code for the complete application.

Now it's time to get to work!

Specifying the Content

If the user selects Currency from the main view in the iPhoneTravel411 application, he or she will see some very basic information about exchange rates, as illustrated in Figure 16-1. Because this information changes rarely (if ever), I'm going to include this information in the application. The way to do this is to include it as a resource, which I explain in a later section. The view the user sees will be the same regardless of whether the user is online or in stored data mode.

If the user selects Weather from the main view, what the user sees *does* depend on whether the device is online or in stored data mode. If the device is online, the user sees a Web page from a Web site with the weather information, as illustrated in Figure 16-2 (left). When in stored data mode, the user gets a message stating that weather data is unavailable when offline, as you can see in Figure 16-2 (right).

If the user selects Heathrow from the main view, he or she sees what's illustrated in Figure 16-3: a view of one particular means of transportation (Heathrow Express, for example) with tabs that will enable her or him to look at the others. When the user is online, the application gets this data off a server on the Internet where I (personally) keep the (up-to-date) information. The only difference between being online and being in stored data mode (offline) is the freshness of the data. In stored data mode, what the user sees is a copy of the information that the application saved the last time the user was online.

If the user selects Map from the main view, he or she sees what's illustrated in Figure 16-3: a map of the destination and some other features I cover in Chapter 17. To see a map, the user must be online (at least in this implementation). When in stored data mode, the user gets a message stating that map data is unavailable when offline.

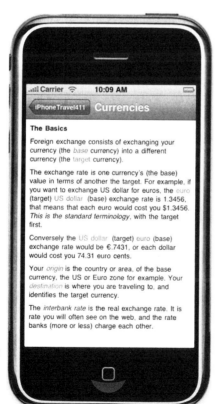

Figure 16-1:
Currency
view.

Figure 16-2:
Weather
views.

In this chapter, I show you how to code the view controllers (Weather
Controller, for example) and their views, as well as the model for each
of the examples shown in Figures 16-1, 16-2, and 16-3. The view controllers
are the key elements here. They're the ones that get the content from the
Destination model object (which you created in the awakeFromNib
method of the RootViewController in Chapter 14) and send it back to the
view to display. With the exception of the MapController (which I explain
in detail in Chapter 17), the views will all be UIWebViews (see Chapter 13
for more on why), and I'll have you creating a unique view controller
(WeatherController, for example) for each and the Destination model
(that interfaces with other model objects) that supports all of them.

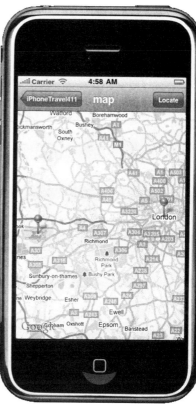

If you had a chance to make your way through Chapter 13, you remember that I said you could use a *generic* view controller and *generic* model objects to display any of the views you're about to create. Well, you can, but I'm not going to recommend doing that in this example for two reasons. It's a lot easier to follow the logic in each of these view controllers and models if I keep them separate, because then I can use more context-specific data names and logic. But more important, although it's not rocket science, what you'd have to do to generalize the behavior of model objects and view controllers involves some work with property lists or databases, and that's not in the cards for this book.

What I *will* do, though, is explain things in such a way that the pattern that underlies it all becomes apparent.

Fair enough?

Creating the View Controller, Nib, and Model Files

Standard operating procedure for iPhone applications is to have a user tap an entry in a main view to get to a subsequent view. For this to work, you need to create a new controller and push it onto the stack. (For more on controllers and stacks as they apply to the iPhoneTravel411 application, see Chapter 14.) To make it happen, you need to code your view controller interface and implementation files, your nib files, and then your model interface and implementation files. (Yes, we're talking a lot of files here — it's all good, so deal with it as best you can.)

Going from abstract to concrete, read on as I spell out what you need to have in place before your users can jump from one view to another in iPhone Travel411 — say, when a user selects the "Heathrow" row in the main view of the app and fully expects to find a new view full of information about (yep) Heathrow airport (although in this example, I illustrate only transportation from and to the airport). This will be the general pattern you can follow for the rest of the rows in this application, or in any application that uses a table view. (For more on table views, check out Chapter 14.)

Adding the controller and nib file

So many files, so little time. Actually, after you get a rhythm going, cranking out the various view controller, nib, and model files necessary to fill your application architecture with content isn't *that* much work. And although I want to start with what happens when the user taps Heathrow, because it allows me to also explain a bit about navigating between views in your program, now is as good a time as any to create all those files you need.

Okay, check out how easy it is to come up with the view controller and nib files:

1. **In the iPhoneTravel411 Project window, select the Classes folder, and then select File⇨New from the main menu (or press ⌘+N) to get the New File window you see in Figure 16-4.**

2. **In the left column of the dialog, select Cocoa Touch Class under the iPhone OS heading, select the UIViewController Subclass template in the top-right pane, and then click Next.**

    Be sure that the With XIB for User Interface check box is also selected.

    You see a new dialog asking for some more information.

3. **Enter** AirportController.m **in the File Name field, as I did in Figure 16-5, and then click Finish.**

The Also create "AirportController.h" checkbox should be checked by default. For our purposes, always check it if it is not.

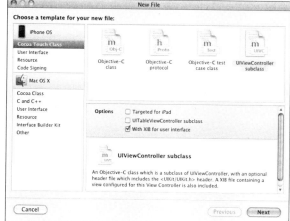

Figure 16-4: Select UIView Controller subclass.

I'm using Airport here instead of Heathrow to get you started thinking more in terms of generic controllers — an airport is an airport after all. What would you have to do to reuse this controller for the other two airports in the main view?

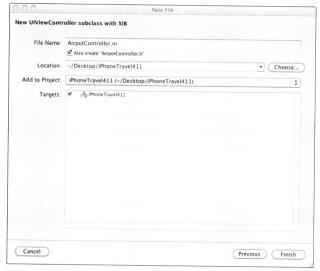

Figure 16-5: Save as Airport Controller.m.

4. **Do the same thing to create a `CityController`, `CurrencyController`, `MapController`, and `WeatherController`.**

   When you're done, you should see `AirportController.h`, `AirportController.m`, `CityController.h`, `CityController.m`, `CurrencyController.h`, `CurrencyController.m`, `MapController.h`, `MapController.m`, `WeatherController.h`, and `WeatherController.m` in your Groups & Files list.

I'm having you set up the `CityController` and nib file and the `City` model object, even though I won't be explaining them in this chapter; however, the code is implemented in the listing on my Web site.

Now you need to do it all over again (and get it out of the way) for the model classes your controllers will use.

Adding the model class

It's a good idea to add a new folder in the Groups & Files list to hold all your new model classes. To do so, select the iPhoneTravel411 Project icon and then choose Project⇔New Group. You'll get a brand-spanking-new folder, named New Group, already selected and waiting for you to type in the name you want. To change what folder a file is in, select and then drag the file to the folder you want it to occupy. The same goes for folders as well. (After all, they can go into other folders.)

1. **Choose File⇔New from the main menu (or press ⌘+N) to get the New File dialog.**

2. **In the leftmost column of the dialog, first select Cocoa Touch Class under the iPhone OS heading, but this time select the Objective-C Class template in the topmost pane, make sure that the drop-down menu Subclass Of has `NSObject` selected. Then click Next.**

   You see a new dialog asking for some more information.

3. **Enter `Destination` in the File Name field and then click Finish.**

4. **Repeat the process to create `.m` and `.h` files for `Airport`, `City`, `Currency`, and `Weather`.**

   Notice that you haven't created a Map class — actually, you don't have to, and I explain why in Chapter 17.

Set up the nib file

For the iPhoneTravel411 application, you want to use a UIWebView to display the airport information you think your users will need. (For the reasoning behind that choice, check out Chapter 13.) You need to set up the UIWeb View by using Interface Builder, but you also need a reference to it from the AirportController so it can pass the content from the model to the view. To do that, you need to create an *outlet* (a special kind of instance that can refer to objects in the nib) in the view controller, just as you did back in Chapter 7 when you were working on the ReturnMeTo application. The outlet reference will be filled in automatically when your application is initialized.

Here's how you should deal with this outlet business (it's the same thing you did in Chapter 7 to set up the ReturnMeToViewController, so if you're a little hazy, you might want to go back and review what you did there):

1. **Within Xcode, add an** airportView (UIWebView) **outlet to the AirportController.h interface file.**

   You declare an outlet by using the keyword IBOutlet in the AirportController interface file.

   ```
   IBOutlet UIWebView    *airportView;
   ```

2. **While you're at it, make the AirportController a UIWebView delegate as well. (You'll need that later.)**

   Here's what it should look like when you are done; changes are in bold:

   ```
   @interface AirportController : UIViewController
                                   <UIWebViewDelegate> {
     IBOutlet UIWebView *airportView;
   ```

3. **Do the File⇨Save thing to save the file.**

   After it's saved — and only then — Interface Builder can find the new outlet.

4. **Use the Groups & Files list on the left in the Project window to drill down to the AirportController.xib file; then double-click the file to launch it in Interface Builder.**

   If the Attributes Inspector window is not open, choose Tools⇨Inspector or press shift+⌘+1. If the View window is not visible, double-click the View icon in the AirportController.xib window.

   If you can't find the AirportController.xib window (you may have minimized it — accidentally, on purpose, whatever) you can get it back by choosing Window⇨AirportController.xib or whichever nib file you're working on.

5. **Select File's Owner in the AirportController.xib window.**

6. **Select `AirportController` from the Class drop-down menu in the Identity Inspector.**

   Doing so (surprise, surprise) tells Interface Builder that the File's Owner is `AirportController`.

7. **Click in the View window and then choose `UIWebView` from the Class drop-down menu in the Identity Inspector.**

   This will change the title of the View window to Web View — but after you have closed and then reopened this file.

8. **Back in the `AirportController.xib` window, right-click File's Owner to call up a contextual menu with a list of connections.**

   You can get the same list by using the Connections tab in the Attributes Inspector.

9. **Drag from the little circle next to the `airportView` outlet in the list onto the Web View window.**

   Doing so connects the `AirportController`'s `airportView` outlet to the Web view.

10. **Go back to that list of connections, and this time drag from the little circle next to the New Referencing Outlet list onto the Web View window.**

11. **With the cursor still in the Web View window, let go of the mouse button.**

    A pop-up menu appears, looking like the one in Figure 16-6.

12. **Choose the Delegate option from the pop-up menu.**

    Doing so makes the `AirportController` the Web view delegate.

    When you're done, the contextual menu should look like Figure 16-7.

If you think about it though, why do you need the `airportView`? There's already a pointer to the `view` object safely nestled in the view controller. There are two reasons.

✔ **I'm lazy.** If I create a second outlet of type `UIWebView`, then every time I access it, I don't have to cast the vanilla Web view into a `UIWebView`, as you can see here:

```
(UIWebView*) [self view] or (UIWebView*) self.view
```

✔ **I'm doing it for you.** It makes the code easier to follow.

Figure 16-6:
Making the
Airport
Controller a
delegate.

Figure 16-7:
Airport
Controller
connections
all in place.

At this point, you have the view controller class set up and you've arranged for the nib loader to create a UIWebView object and set all the outlets for you when the user selects Heathrow for the main view.

Not wanting to be the bearer of bad tidings, although I seem to have developed a skill for that, now do the same thing for CityController, CurrencyController, and WeatherController.

You also need to do the same thing for the MapController, although this time make the View in Step 7 an MKMapView. (It won't need to be a UIWebView delegate, either, but it will need to be a MKMapViewDelegate.)

One thing left to do: You have to add a new framework.

Up until now, all you've needed is the framework that more or less came supplied when you created a project. But now you need a new framework to enable the map view.

1. **Click the disclosure triangle next to Frameworks in the Groups & Files list and then right-click Frameworks.**

2. **From the submenu, select Add and then select Existing Frameworks as I've done in Figure 16-8.**

3. **Select MapKit Framework in the window that appears in Figure 16-9 and then click Add.**

At this point, you have the classes defined for all the view controller and model objects. All that's left for you to do is to enter the code to make it do something — well, maybe not just *something*. How about *exactly* what you want it to do?

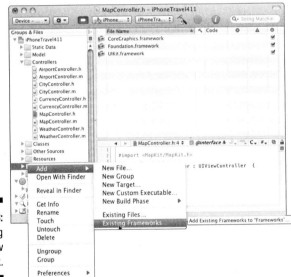

Figure 16-8:
Adding
a new
framework.

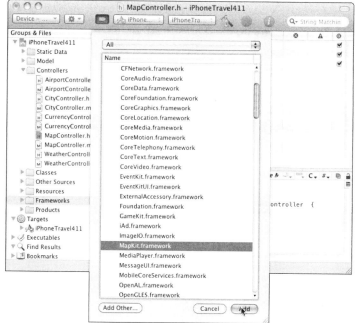

Implementing the View, View Controller, and the Model

You now have all the classes/objects you need to actually implement a map view similar to what you saw earlier on the left side of Figure 16-3.

Check out all the little things you need to do next.

Make sure that the AirportController knows about the objects it needs

Add the following statements to the `AirportController.h` (and all the other controllers you created) file.

```
@class Destination;
```

The compiler needed to know certain things about classes that you were using, such as what methods you defined and so on, and the `#import` statement in the implementation (`.m`) file generally solves that problem. But when you get into objects that point at other objects, you also need to provide that information in the interface file, which can cause a problem if there are circular dependencies (which sounds cool, but I'm not going to get into circular dependencies here; it's beyond the scope of this book). To solve that problem, Objective-C introduces the `@class` keyword. This informs the compiler that `Destination` is a class name. At this point, in the interface file, that is enough for the compiler, but when you actually do use the class (when you create an instance of that class or send it a message, for example), you still have to do the `#import`.

Add the following import statements to the `AirportController.m` file. (You need to add at least `#import "Destination.h"` to all the other controller `.m` files and any other headers files they need as well.)

```
#import "AirportController.h"
#import "Constants.h"
#import "iPhoneTravel411AppDelegate.h"
#import "Destination.h"
```

Initialization

Initialization is one of those nuts and bolts things that's a good idea to pay attention to. It's the way you connect your controller and models (as well as one of the key places to pass the information you'll need if you do create a generic view controller and model). To see how it works, follow along as I walk you through it by using the `AirportController` and model as an example.

It all starts in `RootViewController.m`.

1. **In the `RootViewController`'s `didSelectRowAtIndexPath:` method, create and initialize the view controller that implements the row selected by the user.**

   The following code allocates an Airport controller (a view controller) and then sends it the `initWithDestination:airportID:` message. (I explain this at the very end of Chapter 15; you may want to review that if it has been a while since you looked at it.)

   ```
   targetController = [[AirportController alloc]
           initWithDestination: destination airportID:1];
   ```

   Now on to `AirportController.m`.

2. Start by declaring a new instance variable in `AirportController.h`.

By now, you should know you can put it anywhere between the braces.

```
Destination          *destination;
```

3. Add the `initWithDestination::` method, shown in Listing 16-1, to AirportController.m.

First invoke the superclass's `initWithNibName: bundle:` method:

```
- (id)initWithDestination:(Destination*) aDestination
                     airportID:(int) theAirport {
```

The first thing this method does is invoke its superclass's initialization method. I pass it the nib filename (the one I just created in a previous section) and `nil` as the bundle, telling it to look in the main bundle.

Note that the message to `super` precedes the initialization code added in the method. This sequencing ensures that initialization proceeds in the order of inheritance. Calling the superclass's `initWithNibName: bundle:` method initializes the controller, loads and initializes the objects in the nib file (views and controls, for example), and then sets all its outlet instance variables and Target-Action connections for good measure.

The `init...:` methods all return a pointer to the object created. While not the case here, the reason you assign whatever comes back from an `init...:` method to `self` is that some classes actually return a different class than what you created. The assignment to `self` becomes important if your class is derived from one of those kinds of classes. Keep in mind as well that an `init...:` method can also return `nil` if there's a problem initializing an object. If you're creating an object where that is a possibility, you have to take that into account. (Both of those situations are beyond the scope of this book.)

After the superclass initialization is completed, the `AirportController` is ready to do its own initialization, including saving the `aDestination` argument to the `destination` instance variable.

In Chapter 14, I explain that `Destination` is the model object that the airport controller will get the necessary content from, and here in `initWithDestination::`, `AirportController` saves a reference to it and then sends a message to the `Destination` to get the airport name, which it'll use as a title for the window.

Listing 16-1:   Adding the initWithDestination:airportID: Method

```
- (id)initWithDestination:(Destination*)aDestination
                          airportID:(int) theAirport {

  if (self = [super initWithNibName:@"AirportController"
                             bundle:nil]) {

    destination = aDestination;
    [destination retain];
    self.title = [destination
                       returnAirportName:theAirport];

  }
  return self;
}
```

You have to add a version of this initialization method and the `destination` instance variable to each of your view controllers.

Setting up the view

Your `AirportController` is going to be getting the content for any view you've set up from the `Destination` object — content which it then passes on to the view itself. But before you get a crack at doing that, you need to know how to set up the view. If you refer to Figure 16-3, what you see is `UIWebView`, with a segmented control at the bottom (Train — Taxi — Other, in this example). You use the `ViewDidLoad` method to get your view nice and prepped for its big day. This method was included for you in `AirportController.m` by the `UIViewController` subclass template (albeit, commented out) you chose in the "Adding the controller and nib file" section, earlier in the chapter. Here's the code that was automatically added:

```
/*
// Implement viewDidLoad to do additional setup after
           loading the view, typically from a nib.
- void)viewDidLoad {
  [super viewDidLoad];
}
*/
```

Simply uncomment out this method and follow these steps to add the needed code after `[super viewDidLoad]:`.

While I'm not going to be showing you every detail here, such as adding every instance variable, you'll find all of that and more in the full listing on my Web site.

1. Create and add the Back button.

If you decided to follow Apple's suggestion and not act as a mini browser, you omit this step.

```
UIBarButtonItem *backButton =
                      [[UIBarButtonItem alloc]
                      initWithTitle:@"Back"
                      style:UIBarButtonItemStylePlain
                      target:self
                      action:@selector(goBack:)];
self.navigationItem.rightBarButtonItem = backButton;
[backButton release];
```

In this method, you allocate the button and then assign it to an instance variable that the `UINavigationController` will later use to set up the navigation bar. The `action:@selector(goBack:)` argument is the standard way to specify Target-Action — and is exactly what you did when you created the button that enabled text entry in Chapter 12. It says when the button is tapped, send the `goBack:` message to the `target: self`, which is the `AirportController`. I show you how to implement this shortly.

2. Create the Choice bar to be used for the segmented control at the bottom of the screen.

```
choiceBar = [UIToolbar new];
choiceBar.barStyle = UIBarStyleBlackOpaque;
CGRect viewBounds = self.view.frame;
viewBounds.origin.y = viewBounds.size.height -
      self.navigationController.navigationBar.frame.
                         size.height-kToolbarHeight;
viewBounds.size.height = kToolbarHeight;
[choiceBar setFrame:viewBounds];
```

Here, you're computing a subview's frame by using `viewBounds` and taking into account the height of the Navigation bar. You're also setting the style to `BlackOpaque`, my personal favorite — I hope you don't mind. (You'll find the values for the constants in the `Constants.h` file in the project on my Web site.)

3. Create the segmented control.

```
choiceBarSegmentedControl = [[UISegmentedControl
      alloc] initWithItems: [NSArray
      arrayWithObjects: @"Train", @"Taxi", @"Other",
      nil]];
[choiceBarSegmentedControl addTarget:self action:@
      selector(selectTransportation:) forControlEven
      ts:UIControlEventValueChanged];
```

```
choiceBarSegmentedControl.segmentedControlStyle =
                        UISegmentedControlStyleBar;
choiceBarSegmentedControl.tintColor = [UIColor
                                darkGrayColor];
CGRect segmentedControlFrame = choiceBar.frame;
segmentedControlFrame.size.width =
        choiceBar.frame.size.width - kLeftMargin;
segmentedControlFrame.size.height =
                            kSegControlHeight;
choiceBarSegmentedControl.frame =
                        segmentedControlFrame;
choiceBarSegmentedControl.selectedSegmentIndex = 0;
```

In the first line of code, you're creating a segmented control and an array that specifies the text for each segment. You then set the Target-Action parameters saying that if a segment is tapped by the user (`UIControlEventValueChanged`), then the `selectTransportation::` message is sent to `self`, that is to say the `AirportController`. You then compute the size of the segmented control as you would for any other subview. The last line specifies the initial segment (0) selected when the view is created; before the view is displayed, the `select Transportation::` message is sent to display the content associated with segment 0. (You can see the code for `selectTransportation::` in all its glory in Listing 16-2.)

4. Add the segmented control to the Choice bar.

```
UIBarButtonItem *choiceItem = [[UIBarButtonItem alloc]
        initWithCustomView:choiceBarSegmentedControl];
choiceBar.items =
                [NSArray arrayWithObject:choiceItem];
[choiceItem release];
[choiceBarSegmentedControl release];
```

You get the `choiceBar` (`UIToolbar`) to display controls by creating an array of instances of `UIBarButtonItems` and assigning the array to the `items` property of the `UIToolbar` object (the `choiceBar`). In this case, you create a `UIBarButtonItem` and initialize it with the segmented control you just created. You then create the array and assign it to `items`.

You then can release `choiceItem` because the `NSArray` has a reference to it.

5. Add the Choice bar to the view.

```
[self.view addSubview:choiceBar];
[choiceBar release];
```

You also have to add the following to `Constants.h`:

```
#define kToolbarHeight      48.0
#define kLeftMargin         20.0
#define kSegControlHeight   30.0
```

At this point, you have the view set up, waiting for data, and the segmented control across the bottom that will allow the user to select @"Train", @"Taxi", @"Other".

Responding to the user selection

You've set things up so that when the view is first created — or when the user taps a control — the AirportController's selectTransportation: method is called, allowing the AirportController to hook up what the view needs to display with what the model has to offer. Listing 16-2 shows the necessary code in all its elegance.

Listing 16-2: selectTransportation

```
- (void)selectTransportation:(id) sender {
  [airportView loadRequest:[NSURLRequest requestWithURL:
  [destination returnTransportation:
          (((UISegmentedControl*) sender).
                                  selectedSegmentIndex)]]];
}
```

This is the code that gets executed when the user selects one of the segmented controls (Train, Taxi, Other) that you added to the view — (((UISegmented Control*) sender).selectedSegmentIndex) gives you the segment number. If you'll notice, the controller itself has no idea — nor should it care — what was selected. It just passes what was selected on to the model. That kind of logic should be (and, as you will soon see, *is*) in the model.

All this does is send a message to the model to find out where the data the Web view needs is located, [destination returnTransportation: (((UISegmentedControl*) sender).selectedSegmentIndex)], and then sends a message to the Web view to load it. This is more or less what you did in Chapter 11, but I explain more about the mechanics of this shortly.

Before I show you the code in the model that implements return Transportation:, I want to show you one other thing, and that's the implementation of goBack:, which I specified as the selector when I created the Back button in viewDidLoad. Listing 16-3 has the details.

Listing 16-3: goBack to Where You Once Belonged

```
- (IBAction)goBack:(id) sender {
  if ([airportView canGoBack] == NO )
    [[self navigationController]
                        popViewControllerAnimated:YES];
  else
    [airportView goBack];
}
```

In Chapter 14, I explain how you really need *two* ways to go back from a view. The first way is the left button on the navigation bar, which sends the user back to the previous view controller and its view — the main (table) view.

The second way — a second Back button — is needed in order to return to a "parallel" view, such as when the user touches a link in a view that sent him or her to an external Web page. In the `viewDidLoad` method, I show you how to create that button and I specify there that when the user touches the button, the `goBack:` message should be sent to `AirportController`.

The `UIWebView` actually implements much of the behavior you need here. When the user touches the Back button and this message is sent, you first check with the Web view to see whether there's someplace to go back *to*. (It keeps a backward *and* forward list.) If there's an appropriate retreat, you send the `UIWebView` message (`goBack:`) that will reload the previous page. If not, it means that you're at the Heathrow content page, and you simply "pop" (remove from the stack) the `AirportController` to return to the main window — the same thing the button on the left side of the navigation bar does.

Finally, you need to disable links when you're in stored data mode — if the user isn't online, there's no way to get to the link. `shouldStartLoadWith Request:` is a `UIWebView` delegate method. (Remember, I had you make the `AirportController` a `UIWebView` delegate earlier when you added the `airportView` outlet.) It's called before a Web view begins loading content to see whether you want the load to proceed. I'm only interested in doing something if the user touched a link when he or she is in stored data mode. Listing 16-4 shows the code you need in order to disable links in such a situation. (If you look at the complete listing on my Web site, you'll see you have to do this in the `City` view controller as well.)

Listing 16-4: Disabling Links in Stored Data Mode

```
- (BOOL)webView:(UIWebView *)webView
  shouldStartLoadWithRequest:(NSURLRequest *) request
  navigationType:(UIWebViewNavigationType) navigationType{
    if ((navigationType ==
          UIWebViewNavigationTypeLinkClicked) &&
        ([[NSUserDefaults standardUserDefaults]
              boolForKey:kUseStoredDataPreference])) {
  UIAlertView *alert = [[UIAlertView alloc]
      initWithTitle:@""
      message: NSLocalizedString(@"Link not
              available offline", @"stored data mode")
      delegate:self
      cancelButtonTitle:NSLocalizedString
        (@"Thanks", @"Thanks") otherButtonTitles: nil];
    [alert show];
    [alert release];
```

```
      return NO;
   }
   else {
     if (navigationType ==
           UIWebViewNavigationTypeLinkClicked)
                        airportView.scalesPageToFit = YES;
     else airportView.scalesPageToFit = NO;
     return YES;
   }
}
```

Here's the process the code uses to get the job done for you:

1. **Check to see whether the user has touched an embedded link while in stored data mode.**

```
if ((navigationType ==
        UIWebViewNavigationTypeLinkClicked) &&
   ([[NSUserDefaults standardUserDefaults]
        boolForKey:kUseStoredDataPreference])) {
```

2. **If the user is in stored data mode, alert him or her to the fact that the link is unavailable, and return NO from the method.**

This informs the Web view not to load the link.

```
UIAlertView *alert = [[UIAlertView alloc]
    initWithTitle:@""
    message: NSLocalizedString (@"Link not available
                          offline", @"stored data mode")
    delegate:self
    cancelButtonTitle:NSLocalizedString
       (@"Thanks", @"Thanks") otherButtonTitles: nil];
[alert show];
[alert release];
return NO;
```

You create an alert here with a message telling the user that the link is not available in stored data mode. The Cancel button's text will be @"Thanks".

3. **If you're not in stored data mode, set scalesPageToFit to YES if you're loading a Web page from a link. If you're simply returning to your Web page set it to NO. Finally, return YES to tell the Web view to load from the Internet.**

```
if (navigationType ==
        UIWebViewNavigationTypeLinkClicked)
                     airportView.scalesPageToFit = YES;
else airportView.scalesPageToFit = NO;
return YES;
```

scalesPageToFit is a UIWebView property. If it's set to YES, the Web page is scaled to fit inside of your view and the user can zoom in and out. If it's set to NO, the page is displayed in the view and zooming is disabled.

The reason I haven't just set it to YES — in `viewDidLoad`, for example, as I will in `WeatherController` — is the fact that the page I created in my word processor fits just fine, thank you very much, and I don't want it scalable. You may want to do something else here, of course; I did it this way to show you how (and where) you have control of Web page properties.

If you decided to follow Apple's suggestion and you haven't set up your app to work as a mini browser, you have to disable the links that are available in the content. You can do that in the `shouldStartLoadWithRequest:` method by coding it in the following way:

```
- (BOOL)webView:(UIWebView *) webView
    shouldStartLoadWithRequest:(NSURLRequest *) request
  navigationType: UIWebViewNavigationType)navigationType{

  if (navigationType ==
                    UIWebViewNavigationTypeLinkClicked)
    return NO;
  else return YES;
}
```

If you do decide to go that route, you add this method to `CityController` and `WeatherController` and make `WeatherController` a `UIWebView` delegate as well.

The Destination Model

You're starting to get all your pieces lined up. Now it's time to take a look at what happens when the controller sends messages to the model.

The `Destination` interface, shown in Listing 16-5, shows what methods are available.

Listing 16-5:   Destination.h

```
@class Airport;
@class Currency;
@class Weather;
@class City;

@interface Destination : NSObject {

  Airport    *airport;
  City       *city;
  Currency   *currency;
```

```
    Weather    *weather;
    NSString   *destinationName;

}
- (NSString*) returnAirportName: (int) theAirportID;
- (NSURL*) returnTransportation: (int) aType;
- (NSURL*) returnCityHappenings;
- (NSURL*) returnCurrencyBasics;
- (NSURL*) weatherRealtime;
- (id) initWithName: (NSString*) theDestination;

@end
```

The first method here should look familiar to you because you used it when
you initialized the `AirportController` object in the "Initialization" section,
earlier in the chapter. The next method is invoked from the `select`
`Transportation:` method in `AirportController`.

Now take a look at `returnTransportation:`.

```
- (NSURL*) returnTransportation: (int) aType{
    return [airport returnTransportation: aType];
}
```

Hmm. All this does is turn around and send a message to another model
object, `Airport`. I explain all this indirection in the next section, but for now,
take a look at what goes on in the `Airport` object, as shown in Listing 16-6.

Listing 16-6:   Airport.h

```
@interface Airport : NSObject {

    NSString *airportName;
}

@property (nonatomic, retain)    NSString *airportName;

- (id) initWithName: (NSString*) name airportID: (int)
            theAirport;
- (NSURL*) returnTransportation: (int) transportationType;
- (NSURL*) getAirportData: (NSString*) fileName;
- (void) saveAirportData: (NSString*) fileName
            withDataURL: (NSURL*) url;

@end
```

The second method here should look familiar to you, because it was just
used in the `Destination` method `returnTransportation:`. The last two
are internal methods that are used only by the model itself.

If you're coming from C++, you probably want these last two methods to be private, but there's no private construct in Objective-C. To hide them, you could have moved their declarations to the implementation file and created an Objective-C category. Here's what that would look like:

```
@interface Airport  ()
- (NSURL*) getAirportData: (NSString*) fileName;
- (void) saveAirportData: (NSString*) fileName
                                    withDataURL: (NSURL*) url;

@end
```

In Listing 16-7, you can see the messages sent from `returnTransportation:` as a group.

Listing 16-7:   Airport Model Method Used by Destination

```
- (NSURL*)returnTransportation:(int) transportationType  {

  NSURL *url = [[NSURL alloc] autorelease];

  iPhoneTravel411AppDelegate *appDelegate =
            (iPhoneTravel411AppDelegate *) [[UIApplication
            sharedApplication] delegate];
  BOOL realtime = !appDelegate.useStoredData;
  if (realtime) {
    switch (transportationType) {
      case 0: {
        url = [NSURL URLWithString:
            @"http://nealgoldstein.com/ToFromT100.html"];
        [self saveAirportData:
                    @"ToFromT100.html" withDataURL:url];
        break;
      }
      case 1: {
        url = [NSURL URLWithString:
            @"http://nealgoldstein.com/ToFromT101.html"];
        [self saveAirportData:
                    @"ToFromT101.html" withDataURL:url];
        break;
      }
      case 2: {
        url = [NSURL URLWithString:
            @"http://nealgoldstein.com/ToFromT102.html"];
        [self saveAirportData:
                    @"ToFromT102.html" withDataURL:url];
        break;
      }
    }
  }
```

```
    else  {
      switch (transportationType) {
        case 0:    {
          url = [self getAirportData:@"ToFromT100.html"];
          break;
        }
        case 1: {
          url =   [self getAirportData:@"ToFromT101.html"];
          break;
        }
        case 2: {
          url = [self getAirportData:@"ToFromT102.html"];
          break;
        }
      }
    }
  return url;
}
```

When a message is sent to the model to return the data the view needs to display, it's passed the number of the segmented control that was touched (Train, Taxi, Other). It's the model's responsibility to decide what data is required here.

To concretize all these abstract coding principles a bit, check out how you should deal with the `returnTransportation:` method. The data for each of the choices in the segmented control is on a Web site — `www.nealgold stein.com`, to be precise. First, the method checks to see whether the user is online, or wants to use stored data.

If the user is online, the method constructs the NSURL object that the Web view uses to load the data. (The NSURL object is nothing fancy. To refresh your memory, it's simply an object that includes the utilities necessary for downloading files or other resources from Web and FTP servers or accessing local files.)

```
NSURL *url = [NSURL URLWithString: @"http://nealgoldstein.
        com/ToFromT100.html"];
```

Then the `saveAirportData:` message is sent:

```
[self saveAirportData:@"ToFromT100.html" withDataURL:
        url ];
return url;
```

The `saveAirportData` method in Listing 16-8 downloads and saves the file containing the latest data for whatever transportation (Taxi, for example) the user selected. It's what will be displayed in the current view, and it'll be used later if the user specifies stored data mode.

Listing 16-8: Saving Airport Data

```
- (void) saveAirportData: (NSString*) fileName withDataURL:
                                              (NSURL*) url {

  NSData *dataLoaded = [NSData
                             dataWithContentsOfURL:url];
  if (dataLoaded == NULL)
                      NSLog (@"Data not found %@", url);
  NSArray *paths = NSSearchPathForDirectoriesInDomains
             (NSDocumentDirectory, NSUserDomainMask, YES);
  NSString *documentsDirectory = [paths objectAtIndex:0];
  NSString *filePath = [documentsDirectory
                   stringByAppendingPathComponent:fileName];
  [dataLoaded writeToFile:filePath atomically:YES];
}
```

You did the exact same thing in Chapter 15 when you saved the current state of the application. If you need a refresher here, go back and work through that part of Chapter 15 again.

I've added an NSLog message if the data can't be found. This is a placeholder for error handling that I've left as an exercise for the reader.

This is definitely not the most efficient way to implement saving files for later use, but given the relatively small amount of data involved, the impact is not noticeable. In the MobileTravel411 application, the user has a specific option to download all the data for any city, eliminating going to the Internet twice — once to download and save the data and then again to display the page.

If the user wants stored data to be used, the method returns the stored data as opposed to loading the data for its URL on the Internet. It gets the data by calling the getAirportData: method, which reads the data that was stored in saveAirportData:.

```
return [self getAirportData:@"ToFromT100"];
```

getAirportData: also constructs a NSURL object that the Web view uses to load the data. The NSURL is more than an object that includes the utilities necessary for downloading files from Web and FTP servers. It also works for local files, and in fact, NSURL objects are the preferred way to load the files you'll be interested in.

So you find the path and construct the NSURL object by using that path. This is shown in Listing 16-9.

Listing 16-9:   Getting the Saved Airport Data

```
- (NSURL*)getAirportData:(NSString*) fileName{

    NSArray *paths = NSSearchPathForDirectoriesInDomains
                (NSDocumentDirectory, NSUserDomainMask, YES);
    NSString *documentsDirectory = [paths objectAtIndex:0];
    NSString *filePath = [documentsDirectory
                    stringByAppendingPathComponent:fileName];
    NSURL* theNSURL= [NSURL fileURLWithPath: filePath];
    if (theNSURL == NULL) NSLog (@"Data not there");
    return  theNSURL;
}
```

What's with the Destination Model and All That Indirection?

You have a couple options to create the model objects needed by the view controllers. One way is to have the view controllers create the ones they'll use. For example, the `AirportController` would create the `Airport` object, and so on. That eliminates the indirection you saw in the previous section; you know, having to go through the `Destination` object to the `Airport` object that does the real work.

While this does work, and I've actually done that in past versions, I'd like you to consider a different approach that results in a more extensible program. I explain this in detail in *Objective-C For Dummies,* so if you're curious, you may want to pick up a copy of that book.

One of the advantages of the MVC design pattern I explain in Chapter 2 is that it allows you to separate these three groups in your application (model objects, view objects, and controller objects) and work on them separately. If each group has a well-defined interface, it encapsulates many of the kinds of changes that are often made so that they don't affect the other groups. This is especially true of the model and view controller relationship.

If the view controllers have minimal knowledge about the model, you can change the model objects with minor impact on the view controllers.

As I said, what makes this possible is a well-defined interface, which I show you how to develop in this section. You'll create an interface between the model and the controllers by using a technique called *composition,* which is a useful way to create interfaces.

I'm a big fan of composition, because it's another way to hide what's really going on behind the curtain. It keeps the objects that use the composite object ignorant of the objects the composite object uses and actually makes the components blissfully unaware of each other, allowing you to switch components in and out at will.

The Destination class provides the basis for such an architecture, and while I won't fully implement it here, you'll understand the structure and have no trouble extending it on your own.

```
@class Airport;
@class Currency;
@class Weather;
@class City;

@interface Destination : NSObject {

   Airport    *airport;
   City       *city;
   Currency   *currency;
   Weather    *weather;
   NSString   *destinationName;
}
- (NSString*) returnAirportName: (int) theAirportID;
- (NSURL*) returnTransportation: (int) aType;
- (NSURL*) returnCityHappenings;
- (NSURL*) returnCurrencyBasics;
- (NSURL*) weatherRealtime- (id) initWithName: (NSString*)
           theDestination;

@end
```

Start with what happens when the Destination object is created:

```
@implementation Destination

- (id) initWithName: (NSString*) theDestination {
  if ((self = [super init])) {
    destinationName = theDestination;
    airport = [[Airport alloc] initWithName:
          NSLocalizedString(@"Heathrow", @"Heathrow")
                                            airportID:1];
    currency = [[Currency alloc] initWithCurrency:
          NSLocalizedString(@"pound", @"pound currency")
                                        currencyID: @"GBP"];
    city = [[City alloc] initWithCity:
                NSLocalizedString(@"London", @"London")];
    weather = [[Weather alloc] initWithCity:
                NSLocalizedString(@"London", @"London")];
  }
  return self;
}
```

`Destination` creates the model objects, encapsulating the knowledge of what objects make up the model from the object that creates it (in this case, `Destination` is created by the `RootViewController`). This hides all implementation knowledge from the view controller — all it would know about is the `Destination` object.

So far, so good. Now, how will the various view controllers get the information they need?

```
- (NSURL*)returnTransportation:(int) aType {

  return [airport returnTransportation: aType];
}

- (NSString*)returnAirportName:(int) theAirportID {

  return airport.airportName;
}

- (NSURL*)returnCityHappenings {

  return [city cityHappenings];
}

- (CLLocationCoordinate2D) returnCityLocation {

  return city.coordinate;
}

- (NSString*) returnCityName {

  return city.cityName;
}

- (NSURL*)returnCurrencyBasics {

  return [currency currencyBasics];
}

- (NSURL*)weatherRealtime {

  return [weather weatherRealtime];
}
```

These are the methods that are visible to the view controllers. They also have no idea about the objects that make up the model.

Although it's trivial here and may even appear a bit gratuitous, `Destination` just turns around and essentially resends the message — this architecture becomes important in the more complex applications you'll develop. It'll save you much grief and work as you refactor your code to enhance and extend your app.

The Weather Implementation Model

The Weather view controller, view, and model follow the same pattern put down by the Airport implementation. The controller sends a message to the Destination for some content, which then turns around and sends a message to one of its objects. You can find the code for the Airport, Weather, and Currency (and City) view controllers and models on my Web site, www.nealgoldstein.com.

I want to point out some differences between Weather and the other views that you should be aware of. In the Weather view, there's no segmented control — and no corresponding selectTransportation: message that tells the model what data is needed — so you need to do that in the WeatherController's viewDidLoad method. Listing 16-10 shows how that code looks.

Listing 16-10:   viewDidLoad

```
- (void)viewDidLoad {

  [super viewDidLoad];

  UIBarButtonItem *backButton = [[UIBarButtonItem alloc]
                    initWithTitle: @"Back"
                    style:UIBarButtonItemStylePlain
                    target:self
                    action:@selector(goBack:)] ;
  self.navigationItem.rightBarButtonItem = backButton;
  [backButton release];
  weatherView.scalesPageToFit = YES;
  [weatherView loadRequest:[NSURLRequest requestWithURL:
                    [destination weatherRealtime]]];
}
```

weatherRealtime corresponds to the returnTransportation: method in Airport and constructs and returns the NSURL for the Web view to load. The only other real difference between how Weather and Airport work, as you can see in Listing 16-11, is that there's no data stored for the weather (it seems rather pointless doesn't it?), and you saw that you created an alert for the user if he or she selected Weather in stored data mode. So the Weather object always constructs the NSURL and returns it to Destination.

As I explain earlier, in the "Responding to the user selection," you also set scalesPageToFit to YES. This displays the entire Web page in the view and also allows the user to zoom in and out.

Listing 16-11: **The weatherRealtime Implementation**

```
- (NSURL*)weatherRealtime {

  NSURL *url = [NSURL URLWithString:@"http://www.weather.
          com/outlook/travel/businesstraveler/local/
          UKXX0085?lswe=London,%20UNITED%20KINGDOM&l
          wsa=WeatherLocalUndeclared&from=searchbox_
          typeahead"];
  if (url == NULL) NSLog( @"Data not found %@", url);
  return url;
}
```

That big long URL you see is one I use for weather for London from weather.com. Of course, these things change from time to time, and it may or may not work when you try it. If not, check my Web site for what I'm currently using.

If you don't want the user to be able to click on a link, please refer to the discussion following Step 3 in the "Responding to the user selection" section, earlier in the chapter.

The Currency Implementation Model

The Currency view controller, view, and model follow the same pattern laid down by the Airport and Weather implementations. Again, the complete code is on my Web site at www.nealgoldstein.com. There are only a few differences I need to point out.

Currency is always offline, which means it's a great way to show you how to implement static data. It hardly ever changes, so it works to include it in the application itself.

The content for the Currency view is in a file I created called Currencies.html. To make it available to the application, I need to include it in the application bundle itself, although I could have downloaded it the first time the application ran. (But there's method in my madness. Including it in the bundle does give me the opportunity to show you how to handle this kind of data.)

Now, you can add it to your bundle one of two ways:

✔ Open the Project window and drag an .html file into the Project folder, like you did the icon in Chapter 5.

  It's a good idea to create a new folder within your Project folder as a snug little home for the file. (I named my new folder "Static Data.")

Or

✔ Select Project➪Add to Project and then use the dialog that appears to navigate to and select the file you want. You can see that in Figure 16-10.

The only thing interesting here is that you are going to use some data that you have included with your application as a *resource* (which you can think about as an included file, although it does not "live" in the file system but rather is embedded in the application itself).

In Listing 16-12, you can see that, in the viewDidLoad method in CurrencyController.m, the currencyBasics method sends the message to the view to load the content it gets from Destination.

Listing 16-12:   viewDidLoad

```
- (void)viewDidLoad {

    [super viewDidLoad];
    [currencyView loadRequest:[NSURLRequest requestWithURL:
                     [destination returnCurrencyBasics]]];
}
```

In Listing 16-13, you can see how the currencyBasics method in the Currency.m file constructs the NSURL.

Figure 16-10: Add Currencies. html to the project.

Listing 16-13: currencyBasics method

```
- (NSURL*)currencyBasics {

    NSString *filePath = [[NSBundle mainBundle]
            pathForResource:@"Currencies" ofType:@"html"];
    NSURL* currencyData= [NSURL fileURLWithPath: filePath];
    return currencyData;
}
```

In this case, you're using `pathForResource::`, which is an `NSBundle` method to construct the `NSURL` (You used an `NSBundle` method when you got the application name in the `RootViewController` to set the title on the main window back in Chapter 14.) Just give `pathForResource::` the name and the file type.

Be sure that you provide the right file type; otherwise, this technique won't work.

Notice the Pattern

In Chapter 13, I claim there was a pattern that you could recognize that would enable you to create generic view controllers and some of the model objects. Because by this time your head may be ready to explode, let me give you some direction.

If this controller coding is beginning to look the same, that's because it is — in order to load content into Web views, you create an `NSURL` with the right Web URL, file path, or resource path, and you're off to the races. All the view controllers get an `NSURL` from the model and then send a message to the Web view to load the data (the `loadRequest:` message). The model objects all construct the `NSURL`. Of course, in all my examples, the requests are hard-coded, but in your application, it can be constructed by using a plist or database (both being beyond the scope of this book).

As I said, it becomes somewhat pedestrian, but then again, think about how this architecture enables you to easily add new content views.

What's Next?

If you compile the code posted on my Web site up to Chapter 16, you can see what's on the left side of Figure 16-11. If you click or touch one of the rows, you get the kinds of content I specified (courtesy of some files I have on my Web site). What's interesting, though, is that if you select Map, you get what you see on the right side of Figure 16-11. No, it's not that this was a long road and you forgot what you did to get that. In fact, you did nothing but set up the nib file, and that's the beauty of the map framework. (In the interest of full disclosure, I also added an initialization method consistent with all the other view controller initialization methods, but that made no difference.) In Chapter 17, I expand on the map framework and explain a little about user location as well. As I mention, this is one of the key features of the iPhone and enables applications to provide not only context-based information (information about what is going on in the context) as well as functionality, but also information about the context itself.

Figure 16-11:
You get a
map for
free.

Chapter 17

Finding Your Way

In This Chapter

▶ Using the map framework

▶ Specifying the location and zoom level of a map

▶ Annotating significant locations on the map

▶ Identifying the iPhone's current location

*U*p until now, the functionality I've been focusing on is the kind that enables you to do what you need to *in a context,* like get from the airport to your hotel, change money, tip appropriately, and so on. But there's also another side to context: the fact that the iPhone can provide information *about the context.* Weather is one example, and a map is another.

As the iPhone has evolved over the last few years, Apple, as well as app developers, have put a lot of emphasis on location-based services. These kinds of services allow you to orient a user to where he or she is physically, as well as stay connected to other people who are nearby.

Personally, I'm a big fan of these kinds of services, and if your user can use them, I strongly suggest making them a part of your application.

Keep the following in mind: On the iPhone, location-based services have two distinct aspects:

✔ **Mapping:** Provided by the Map Kit framework, mapping enables you to easily display a map, with annotations, very similar to the Map application.

✔ **Core Location:** Provided by the Core Location framework, Core Location includes a set of Objective-C services capable of accessing a user's location and heading changes from the onboard compass. (*Heading* — the compass direction that the iPhone is pointed — is available on the 3GS and above.)

In this chapter, I take you through embedding maps into your application. In Chapter 18, I explain Core Location and how you can even use Core Location services in the background — enabling you to send updates to a user through local notifications.

Mapping the Light Fantastic

I figured out that, as people began to realize the kinds of solutions the iPhone could deliver, I would be doing potential users of my MobileTravle411 app a disservice if I didn't include in the app the ability to display a map. To many travelers, nothing brands you more as a tourist than unfolding a large map (except, of course, looking through a thick guidebook). In this chapter, I show you how to take advantage of the iPhone's built-in capability to display a map of virtually anywhere in the world, as well as determine your location and then indicate it on the map. As I mention earlier, its awareness of your location is one of the things that enables you to develop a totally new kind of application and really differentiate an iPhone application from one found on a desktop.

For that reason, in this chapter I return to the (more or less) step-by-step format I used in showing you how to build the ReturnMeTo application. Being able to build maps into your application is an important new feature in the iPhone 3.0 SDK and beyond, and I want to be sure that you really understand how to use it. So, if you want to follow along with me, you can start with the project in the folder named "iPhoneTravel411 Chapter 16" found on my Web site (www.nealgoldstein.com). The final version of what you end up with can then be found in the folder "iPhoneTravel411 Chapter 17."

Oh, and by the way, it turns out that working with maps is one of the most fun things you can do on the iPhone because Apple makes it so easy. In fact, you saw this in Chapter 16, where you displayed a map (which in fact does actually support the standard panning and zooming gestures) by simply creating a view controller and a nib file.

In this chapter, you add a lot to that map. I show you how to center your map on an area you want to display (Heathrow airport or London, for example), add annotations (those cute pins in the map that display a callout to describe that location when you touch them), and even show the user's current location. (Although I don't cover it here, you can also turn the iPhone's current address into an Address Book contact.)

Figure 17-1 shows a better way than the standard map view to find your way from the airport.

One of the great features of iPhone 3.0 SDK and beyond is a new framework: MapKit. As explained in Chapter 16, MapKit enables you to bring up a simple map, and also do things with your map without having to do much work at all.

The map looks like the maps in the built-in applications and creates a seamless mapping experience across multiple applications.

Figure 17-1:
Heathrow
to London
map.

MKMapView

The essence of mapping on the iPhone is the MKMapView. It is a UIView subclass, and you can use it out of the box to create a world map. You use this class as-is to display map information and to manipulate the map contents from your application. It enables you to center the map on a given coordinate, specify the size of the area you want to display, and annotate the map with custom information.

You added the MapKit framework to the project in Chapter 16.

When you initialize a map view, you can specify the initial region for that map to display. You do this by setting the *region* property of the map. A region is defined by a center point and a horizontal and vertical distance, referred to as the *span*. The span defines how much of the map will be visible and results in a zoom level. The smaller the span, the greater the zoom.

The map view supports the standard map gestures.

✔ Scroll

✔ Pinch zoom

✔ Double-tap zoom in

✔ Two-finger-tap zoom out (You might not even know about that one.)

You can also specify the map type — regular, satellite, or hybrid — by changing a single property.

Because MapKit was written from scratch, it was developed with the limitations of the iPhone in mind. As a result, it optimizes performance on the iPhone by caching data as well as managing memory and seamlessly handling connectivity changes (like moving from 3G to Wi-Fi).

The map data itself is Google-hosted map data, and network connectivity is required. And because the MapKit framework uses Google services to provide map data, using it binds you to the Google Maps/Google Earth API terms of service.

Although you shouldn't subclass the MKMapView class itself, you can tailor a map view's behavior by providing a delegate object. The delegate object can be any object in your application as long as it conforms to the MKMapViewDelegate protocol, and as you'll see, you can easily use the model you developed in Chapter 16 to do that. (I wasn't kidding when I said you were done with the tedious creation of files!)

Enhancing the Map

Having this nice global map centered on the United States is kind of interesting, but not very useful if you are planning to go to London. The following sections show you what you have to do to make the map more useful, as well as center the map on Heathrow and London.

In Chapter 16, you created a view controller class (MapController) with the right outlets and a nib file that creates a MKMapView object and sets all the outlets for you when the user selects Map from the main view. You can also have an initialization method set up.

Adding landscape mode and the current location

To start this enhancement business off with a bang, it would be very useful to be able to see any map in landscape mode.

Go back to Xcode and add the following method to MapController.m:

```
- (BOOL)shouldAutorotateToInterfaceOrientation:
        (UIInterfaceOrientation)toInterfaceOrientation {

  return YES;
}
```

That's all you have to do to view the map in landscape mode — and MapKit will take care of it for you! (This is starting to be real fun.)

What about showing your location on the map? That's just as easy!

Although you added the `MapKit` framework in Chapter 16, you still have to tell `MapController` to use it. Add the following to `MapController.h`:

```
#import <MapKit/MapKit.h>
```

In the `MapController.m` file, uncomment out `viewDidLoad` and add the code in bold.

```
- (void)viewDidLoad {
    [super viewDidLoad];

    mapView.showsUserLocation = YES;
}
```

`showsUserLocation` is a `UKMapView` property that tells the map view whether to show the user location. If set to `YES`, you get that same blue pulsing dot displayed in the built-in Map application.

If you were to compile and run the application as it stands, you'd get what you see in Figure 17-2 — a map of the USA in landscape mode with a blue dot that represents the phone's current location. (There may be a lag until the iPhone is able to determine that location, but you should see it eventually.) Of course, to see it in landscape mode, you have to turn the iPhone, or choose Hardware⇨Rotate Right (or Rotate Left) from the Simulator menu, or press ⌘+right (or left) arrow.

If you don't see the current location, you may want to check and make sure that you've connected the `mapView` outlet to the map view in the nib file. You did this back in Chapter 16.

That is the current location *if you are running on the iPhone.* If you're running on the Simulator, that location is Apple. Touching the blue dot also displays what is called an *annotation,* and you find out how to customize the text to display whatever you cleverly come up with, including — as explained in the "Annotations" section, later in this chapter — the address of the current location.

Figure 17-2:
Displaying a map in landscape mode with a user location.

It's about the region

While this map is cute, it's still not that useful for our purposes.

As I state at the beginning of this chapter, ideally when you land at Heathrow (or wherever), you should see a map that centers on Heathrow as opposed to the United States. To get there from here, however, is also pretty easy.

First look at how you center the map.

Add the following code to `MapController.m`:

```
- (void)updateRegionLatitude:(float) latitude
        longitude:(float) longitude
        latitudeDelta:(float) latitudeDelta
          longitudeDelta:(float) longitudeDelta {
  MKCoordinateRegion region;
  region.center.latitude = latitude;
  region.center.longitude = longitude;
  region.span.latitudeDelta = latitudeDelta;
  region.span.longitudeDelta = longitudeDelta;
  [mapView setRegion:region animated:NO];
}
```

Also add the method declaration to `MapController.h`:

```
- (void)updateRegionLatitude:(float)
      latitude longitude:(float)
      longitude latitudeDelta:(float)
        latitudeDelta longitudeDelta:(float) longitudeDelta;
```

Setting the *region* is how you center the map and set the zoom level. All of this is accomplished by the statement

```
[mapView setRegion:region animated:NO];
```

A *region* is a map view property that specifies four things (as illustrated in Figure 17-3).

1. `region.center.latitude` specifies the latitude of the center of the map.

2. `region.center.longitude` specifies the longitude of the center of the map.

   For example, if I were to set those values as

   ```
   region.center.latitude = 51.471184;
   region.center.longitude = -0.452542;
   ```

   the center of the map would be at Heathrow airport.

3. `region.span.latitudeDelta` specifies the north-to-south distance (in degrees) to display on the map. One degree of latitude is always approximately 111 kilometers (69 miles). A `region.span.latitude Delta` of 0.0036 specifies a north-to-south distance on the map of about a quarter of a mile. Latitudes north of the equator have positive values, and latitudes south of the equator have negative values.

4. `region.span.longitudeDelta` specifies the east-to-west distance (in degrees) to display on the map. Unfortunately, the number of miles in one degree of longitude varies based on the latitude. For example, one degree of longitude is approximately 69 miles at the equator but shrinks to 0 miles at the poles. Longitudes east of the zero meridian (by international convention, the zero or Prime Meridian passes through the Royal Observatory, Greenwich, in east London) have positive values, and longitudes west of the zero meridian have negative values.

While the span values provide an implicit zoom value for the map, the actual region you see displayed may not equal the span you specify because the map will go to the zoom level that best fits the region that's set. This also means that even if you just change the center coordinate in the map, the zoom level may change because distances represented by a span change at different latitudes and longitudes. To account for that, those smart developers at Apple included a property you can set that will change the center coordinate without changing the zoom level.

```
@property (nonatomic) CLLocationCoordinate2D
          centerCoordinate
```

When you change this property, the map is centered on the new coordinate and updates span values to maintain the current zoom level.

That `CLLocationCoordinate2D` type is something you'll be using a lot, so I'd like to explain that before I take you any farther.

`CLLocationCoordinate2D` type is a structure that contains a geographical coordinate using the WGS 84 reference frame (the reference coordinate system used by the Global Positioning System).

```
typedef struct {
CLLocationDegrees latitude;
CLLocationDegrees longitude;
} CLLocationCoordinate2D;
```

✔ `latitude` is the latitude in degrees. This is the value you set in the code you just entered (`region.center.latitude = latitude;`).

✔ `longitude` is the longitude in degrees. This is the value you set in the code you just entered (`region.center.longitude = longitude;`).

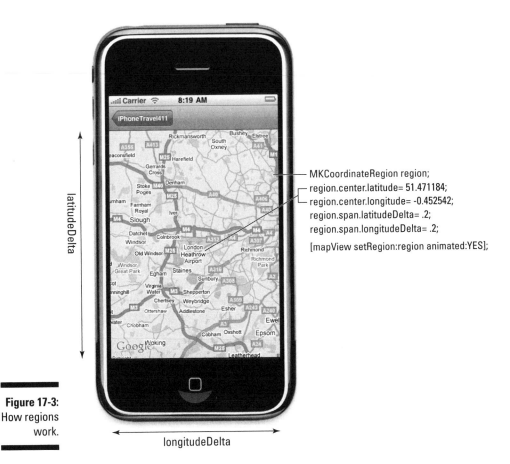

latitudeDelta

MKCoordinateRegion region;
region.center.latitude= 51.471184;
region.center.longitude= -0.452542;
region.span.latitudeDelta= .2;
region.span.longitudeDelta= .2;

[mapView setRegion:region animated:YES];

longitudeDelta

Figure 17-3:
How regions
work.

To center the map display on Heathrow, you send the `updateRegion`
`Latitude:longitude:` `latitudeDelta:longitudeDelta` message
(the code you just entered) when the view is loaded — that is, in the
`viewDidLoad:` method. You already added some code there to display the
current location, so add the code in bold.

```
- (void)viewDidLoad {

    [super viewDidLoad];
    mapView.showsUserLocation = YES;
    CLLocationCoordinate2D initialCoordinate =
                            [destination initialCoordinate];
    [self updateRegionLatitude:
        initialCoordinate.latitude
        longitude: initialCoordinate.longitude
        latitudeDelta:.2 longitudeDelta:.2];
    self.title = [destination mapTitle];
}
```

Here's what this code does:

1. The `initialCoordinate` message is sent to the `Destination` object (remember your model from Chapter 16) to get the initial coordinates you want displayed. You're adding some more functionality to the model, whose responsibly now includes specifying that location. The user may have requested that location when he or she set up the trip (I don't cover that topic in this book, leaving it as an exercise for the reader), or it may have been a default location that you decided on when you wrote the code (an airport specified in the destination, for example).

2. Sets the map title by sending the `mapTitle` message to the `Destination` object — adding another model responsibility.

For all this to work, of course, you have to add the following code to `Destination.m`. The following code will return the latitude and longitude for Heathrow:

```
- (CLLocationCoordinate2D) initialCoordinate {

    CLLocationCoordinate2D startCoordinate;
    startCoordinate.latitude=51.471184;
    startCoordinate.longitude=-0.452542;
    return startCoordinate;
}
- (NSString*) mapTitle{

    return @" map";
}
```

You have to include the `MapKit` in `Destination`, so add the following to `Destination.h`:

```
#import <MapKit/MapKit.h>
```

You also have to add the following to `Destination.h` (right there after the braces):

```
- (CLLocationCoordinate2D) initialCoordinate;
- (NSString*)mapTitle;
```

If you compile and build your project, you should see what's shown in Figure 17-4.

At this point, when the user touches Map in the main view, iPhoneTravel411 displays a map centered on Heathrow, and if you pan (a tedious task you'll fix soon) over to Cupertino (or wherever you are), you see the blue dot.

If you tap the dot, shown back in Figure 17-2, you see a callout known as an *annotation* displaying "Current Location." You can also add annotations on your own, which is what you do in the next section.

Figure 17-4: Regions determine what you see on the map.

Tracking location changes

You can also track changes in user location by using key-value observing, which enables you to move the map as the user changes location. I don't go into detail on key-value observing other than to show you the code. It's very similar to when you registered for the `UIKeyboardWillShowNotification` in the ReturnMeTo application in Chapter 8.

First, you add the code in bold to `viewDidLoad:` in `MapController.m` to add an observer that's to be called when a certain value is changed — in this case `userLocation`.

```
- (void)viewDidLoad {
  [super viewDidLoad];
  mapView.showsUserLocation = YES;
  CLLocationCoordinate2D initialCoordinate =
                                          [map
  initialCoordinate];
  [self updateRegionLatitude: initialCoordinate.latitude
    longitude:
                 initialCoordinate.longitude
                 latitudeDelta:.2 longitudeDelta:.2];

  self.title = [destination mapTitle];
  [mapView.userLocation addObserver:self
    forKeyPath:@"location"
                                          options:0

  context:NULL];
}
```

Adding that code causes the `observeValueForKeyPath::` message to be sent to the observer (self or the `RootViewController`). To implement the method in `Destination.m`, enter the following:

```
-  (void)observeValueForKeyPath:(NSString *) keyPath
                  ofObject:(id)object change:(NSDictionary *)
     change
                  context:(void *) context {

   NSLog (@"Location changed");
}
```

In this method, the `keyPath` field returns `mapView.userLocation.location`, which you can use to get the current location. In this example, I'm simply displaying a message on the Debugger Console, but as I said, after the user moves a certain amount, you may want to re-center the map.

Technical Stuff: This is not exactly the same location you'd get from `CLLocationManager` — it's optimized for the map, while `CLLocationManager` (that you will use in Chapter 18) provides the raw user location.

Of course, you have to run this on the iPhone for the location to change.

Annotations

The `MKMapView` class supports the ability to annotate the map with custom information. There are two parts to the annotation — the annotation itself, which contains the data for the annotation, and the annotation view that displays the data.

The annotation proper

An annotation plays a similar role to the dictionary you created in Chapter 16 to hold the text to be displayed in the cell of a table view. Both of them act as a model for their corresponding view, with a view controller connecting the two.

Annotation objects are any object that conforms to the `MKAnnotation` protocol; typically, they are existing classes in your application's model. The job of an annotation object is to know its location (coordinate) on the map along with the text to be displayed in the callout. The `MKAnnotation` protocol requires a class that adopts that protocol to implement the `coordinate` property. In this case, it makes sense for the `Airport` and `City` model objects to add the responsibilities of an annotation object to their bag of tricks. After all, the `Airport` and `City` model objects already know what airport or city they represent, respectively. It makes sense for these objects to have the coordinate and callout data as well.

Here's what you need to do to make that happen:

1. **Add the MapKit import statement to the Airport.h and City.h files.**

   ```
   #import <MapKit/MapKit.h>
   ```

2. **Have the City and Airport classes adopt the MKAnnotation protocol.**

   ```
   @interface City : NSObject <MKAnnotation> {
   @interface Airport : NSObject <MKAnnotation> {
   ```

3. **Add the following instance variables to both the Airport.h and City.h files.**

   ```
   CLLocationCoordinate2D coordinate;
   ```

4. **Add the following property and method to both the Airport.h and City.h files.**

   ```
   @property (nonatomic) CLLocationCoordinate2D
           coordinate;
   - (NSString*) title;
   ```

   The MKAnnotation protocol requires a coordinate property — the title method is optional.

5. **Add a synthesize statement to both the Airport.m and City.m files.**

   ```
   @synthesize coordinate;
   ```

6. **Implement the Airport's title method by adding the following to the Airport.m file.**

   ```
   - (NSString*)title{

     return airportName;
   }
   ```

   Airport provides the airport name for the callout text.

7. **Implement the City's title method by adding the following to the City.m file.**

   ```
   - (NSString*)title{

     return cityName;
   }
   ```

   City provides the city name for the callout text.

8. **Add the code in bold to the initWithName:: method in Airport.m to provide the coordinate data for Heathrow.**

   ```
   - (id)initWithName:(NSString*) name airportID:(int)
                                          theAirport{

     if ((self = [super init])) {
       airportName = name;
   ```

```
        coordinate.latitude = 51.471184;
        coordinate.longitude= -0.452542;
    }
    return self;
}
```

`Airport` is assigning the latitude and longitude of Heathrow to the `coordinate` property, which will be used by the map view to position the annotation.

9. **Add the code in bold to the `initWithCity:` method in `City.m` to provide the coordinate data for London.**

```
- (id)initWithCity:(NSString*) name {
    if ((self = [super init])) {
        self.cityName = name;
        coordinate.latitude = 51.500153;
        coordinate.longitude= -0.126236;
    }
    return self;
}
```

`City` is assigning the latitude and longitude of London to the `coordinate` property, which will be used by the map view to position the annotation.

10. **Add the code in bold to `Destination.m`.**

```
@synthesize annotations;

- (id)initWithName:(NSString*) theDestination {
    if ((self = [super init])) {
        destinationName = theDestination;
        airport = [[Airport alloc] initWithName:
                NSLocalizedString(@"Heathrow", @"Heathrow")
                                        airportID:1];
        currency = [[Currency alloc] initWithCurrency:
            NSLocalizedString(@"pound", @"pound currency")
                                    currencyID: @"GBP"];
        city = [[City alloc] initWithCity:
                NSLocalizedString(@"London", @"London")];
        weather = [[Weather alloc] initWithCity:
                NSLocalizedString(@"London", @"London")];
        annotations = [[NSMutableArray alloc]
                                    initWithCapacity:4];
        [annotations addObject:airport];
        [annotations addObject:city];
    }
    return self;
}
```

The `Destination` object creates an array of annotation objects. (I show you how it's used next.)

11. Add the annotations array instance variable and make the annotations array a property by adding the following to Destination.h.

```
NSMutableArray *annotations;
...
@property (nonatomic, retain)
                          NSMutableArray * annotations;
```

So far, so good. You have two objects, City and Airport that have adopted the MKAnnotation protocol, declared a coordinate property, and implemented a title method. The Destination object then created an array of these annotations. The only thing left to do is send the array to the map view to get the annotations displayed. As you recall, in Chapter 16 this was how you managed to have buttons displayed in the toolbar at the bottom of the airport view.

Displaying annotations

Displaying the annotations is easy. All you have to do is add the line of code in bold to MapController.m.

```
- (void)viewDidLoad {
  [super viewDidLoad];
  mapView.showsUserLocation = YES;
  CLLocationCoordinate2D initialCoordinate = [destination
          initialCoordinate];
  [self updateRegionLatitude: initialCoordinate.latitude
          longitude:initialCoordinate.longitude
          latitudeDelta:.2 longitudeDelta:.2];
  self.title = [destination mapTitle];

  [mapView.userLocation addObserver:self
                          forKeyPath:@"location" options:0
                          context:NULL];
  [mapView addAnnotations: destination.annotations];
}
```

The MapController sends the addAnnotations: message to the map view, passing it an array of objects that conform to the MKAnnotation protocol; that is, each one has a coordinate property and an optional title (and subtitle) method if you want to display something in the annotation callout.

The map view places annotations on the screen by sending its delegate the mapView:viewForAnnotation: message. This message is sent for each annotation object in the array. Here you can create a custom view, or return `nil` to use the default view. (If you don't implement this delegate method — which you won't — the default view is also used.)

Creating custom annotation views is beyond the scope of this book (although I'll tell you that the most efficient way to provide the content for an annotation view is to set its image property). For the time being, the default annotation view is fine for your purposes. It displays a pin in the location specified in the coordinate property of the annotation delegate (`City` and `Airport`, in this case), and when the user touches the pin, the optional title and subtitle text displays if the `title` and `subtitle` methods are implemented in the annotation delegate.

This default view displays red pins stuck in a map. Later, I show you how to use a built-in annotation view that's a bit more flexible. It allows you to specify a pin color, animate the pins so they drop onto the map, and change some properties so the pin becomes draggable.

You can also add callouts to the annotation callout, such as a Detail Disclosure button (the one that looks like a white chevron in a blue button in a table-view cell), or the Info button (like the one you see in many of the utility apps) without creating your own annotation view. Again, another exercise for you to tackle on your own, if you like.

If you compile and build your project, you can check out one of the annotations you just added in Figure 17-5.

Figure 17-5:
An
annotation.

Displaying multiple annotations

Although what you see in the map view is good, under some circumstances (like when you just arrived at the airport, for example) it would really be better if the user could automatically see both Heathrow and London (and their respective annotations) on the map without hard-coding the region as you have been doing so far. This isn't that difficult to do. Add the code in Listing 17-1 to `MapController.m` and the declaration to `MapController.h`.

Listing 17-1:   regionForAnnotationGroup

```
- (MKCoordinateRegion)
              regionForAnnotationGroup: (NSArray*) group {

    double maxLonWest= 0;
    double minLonEast = 180;
    double maxLatNorth = 0;
    double minLatSouth = 180;

    for (City* location in group) {
      if (fabs(location.coordinate.longitude ) >
            fabs(maxLonWest))
          maxLonWest = location.coordinate.longitude;
      if (fabs(location.coordinate.longitude ) <
            fabs(minLonEast))
          minLonEast = location.coordinate.longitude;
      if (fabs(location.coordinate.latitude ) >
            fabs(maxLatNorth))
          maxLatNorth = location.coordinate.latitude;
      if (fabs(location.coordinate.latitude ) <
            fabs(minLatSouth))
          minLatSouth = location.coordinate.latitude;
    }

    double centerLatitide =
         maxLatNorth - (((maxLatNorth) - (minLatSouth))/2);
    double centerLongitude =
         maxLonWest - (((maxLonWest) - (minLonEast))/2);

    MKCoordinateRegion region;
    region.center.latitude =  centerLatitide;
    region.center.longitude = centerLongitude;
    region.span.latitudeDelta =
                  fabs(maxLatNorth - minLatSouth)+.005;
    if (fabs(maxLatNorth - minLatSouth) <= .005) region.
          span.latitudeDelta = .01;
    region.span.longitudeDelta =
                  fabs(maxLonWest - minLonEast)+.005;
```

```
    if (fabs(maxLonWest - minLonEast) <= .005) region.span.
            longitudeDelta = .01;

    return region;
}
```

This code computes the region that includes both annotations. Frankly, this is really beyond the scope of this book, but it's handy to be able to know how to do this, so I include here. I just summarize how it works and leave it for you to go through it step by step.

In general terms, here's how the code works: You want both annotations to fit on one screen, so the code goes through each annotation and determines the maximum north and south latitudes and the maximum east and west longitudes. The only trick here is that, because latitude and longitude can be negative, it uses a function `fabs` to get the absolute value of a floating point number. After that, it's simply a matter of finding the center latitude and longitude and setting the region `center` and then taking the maximum west and maximum east longitude and the maximum north and maximum south latitude and using that as the `span`. (I also decrease the span by .005 in both directions to make sure that no pins are right on the edge of the screen.) I decided that I didn't want the `span` to ever be less than `.005`, so if it is, I arbitrarily make it `.01`.

All that is left is to change the `MapController viewDidLoad` method to use this method rather than the `Destination initialCoordinate` and `updateRegionLatitude:::::`.

To do that, add the stuff in bold and then delete the stuff in strikethrough in `viewDidLoad` as shown in Listing 17-2.

Listing 17-2:   viewDidLoad Now Results in the Annotations Being Visible

```
- (void)viewDidLoad {

    [super viewDidLoad];
    mapView.showsUserLocation = YES;
//  CLLocationCoordinate2D initialCoordinate =
//  [destination initialCoordinate];
//  [self updateRegionLatitude: initialCoordinate.latitude
            longitude: initialCoordinate.longitude
//               latitudeDelta:.2 longitudeDelta:.2];
    self.title = [destination mapTitle];
    [mapView.userLocation addObserver:self
            forKeyPath:@"location" options:0 context:NULL];
    [mapView addAnnotations:
```

(continued)

Listing 17-2 *(continued)*

```
                                    [destination createAnnotations]];
    NSArray* annotations = [destination createAnnotations];
    [mapView addAnnotations:annotations];
    [mapView setRegion:[self regionForAnnotationGroup:
            annotations] animated:NO];
}
```

You also have to import `City.h`. You can see the results in Figure 17-6.

At this point you can also delete `initialCoordinate` in `Destination.h` and `.m` and `updateRegionLatitude` in `MapController.h` and `.m`.

Figure 17-6:
A better
way to
compute a
region.

After you make regions something you compute rather than something you hard-code, all of a sudden you're confronted with another issue, and that is: You want the map to be able to be displayed in both landscape and portrait mode. The problem is that if the application opens in landscape mode and computes the correct region, then when you switch to portrait mode, the annotations are no longer visible.

So this is a good time to introduce you to `didRotateFromInterface Orientation`. This message is sent to the `MapController` when the user inference orientation changes. Fortunately, all you have to do to get this to work for you is add the code in Listing 17-3 to `MapController.m`.

Listing 17-3: Accounting for User Rotation Changes

```
- (void)didRotateFromInterfaceOrientation:
        (UIInterfaceOrientation)fromInterfaceOrientation {

[mapView setRegion:
    [self regionForAnnotationGroup:destination.annotations]
                                        animated:NO];
}
```

All you need to do is set the new region size. You do that in `regionFor AnnotationGroup:animated:`, and the map does the rest for you.

Going to the Current Location

Although you can pan to the user location on the map, in this case, it's kind of annoying, unless you're actually coding this at or around London or Heathrow. To remove at least that annoyance from your life, I want to show you how easy it is to add a button to the navigation bar to zoom you in to the current location and then back to the map region and span you're currently displaying.

1. **Add the following code to add the button in the `MapController` method `viewDidLoad`.**

   You have quite a bit of code there, so this is just what to add.

   ```
   UIBarButtonItem *locateButton =
       [[UIBarButtonItem alloc] initWithTitle: @"Locate"
       style:UIBarButtonItemStylePlain target:self
       action:@selector(goToLocation:)];
   self.navigationItem.rightBarButtonItem = locateButton;
   [locateButton release];
   ```

   If this looks familiar, it's because this is what you did to add the Back button in Chapter 16. When the user taps the button, you've specified that the `goToLocation:` message is to be sent (`action:@selector(goToLocation:)`) to the `MapController` (`target:self`).

2. **Add the `goToLocation:` method to `MapController.m`.**

   ```
   - (IBAction)goToLocation:(id)sender{
     MKUserLocation *annotation = mapView.userLocation;
     CLLocation *location = annotation.location;
     if (nil == location)
       return;
     CLLocationDistance distance =
               MAX(4*location.horizontalAccuracy,500);
     MKCoordinateRegion region =
             MKCoordinateRegionMakeWithDistance
             (location.coordinate, distance, distance);
   [mapView setRegion:region animated:NO];

     self.navigationItem.rightBarButtonItem.action =
                                 @selector(goToTrip:);
     self.navigationItem.rightBarButtonItem.title =
                                 @"Map";
   }
   ```

When the user presses the Locate button, you first check to see whether the location is available. (Note that it may take a few seconds after you start the application for the location to become available.) If not, you simply return. (You could, of course, show an alert informing the user what is going on and to try again in ten seconds or so — I leave that up to you.)

If the location is available, you compute the span for the region you'll be moving to. In this case, the code

```
CLLocationDistance distance =
            MAX(4*location.horizontalAccuracy,1000);
```

computes the span to be four times the horizontalAccuracy of the device (but no less than 1000 meters). horizontalAccuracy is a radius of uncertainty given the accuracy of the device; that is, the user is somewhere within that circle.

You then call the MKCoordinateRegionMakeWithDistance function that creates a new MKCoordinateRegion from the specified coordinate and distance values. distance and distance correspond to latitudinalMeters and longitudinalMeters, respectively.

If you didn't want to change the span, you could have simply set the map view's centerCoordinate property to userLocation, and, as I said in the "It's about the region" section, earlier in this chapter, that would have centered the region at the userLocation coordinate without changing the span.

3. **Change the title on the button to "Map," and the @selector to (goToTrip:), which means that the next time the user touches the button, the goToTrip: message will be sent, so you'd better add the following code:**

```
- (IBAction) goToTrip:(id)sender{

    [mapView setRegion:[self regionForAnnotationGroup:
            destination.annotations] animated:NO];
    self.navigationItem.rightBarButtonItem.title =
                                        @"Locate";
    self.navigationItem.rightBarButtonItem.action =
                            @selector(goToLocation:);
}
```

But there's a bit of a problem here. In Listing 17-3, I set the region if the user rotated the device into landscape mode by using regionForAnnotation Group:. The problem is that it will display the region including Heathrow and London.

So what I want to do is not set the region if I'm showing the user location and let the view controller implement the default behavior. Although you could have used an instance variable to let you know when you were displaying the user location, I take the easy way out, as you can see in Listing 17-4, and check to see what the button says instead.

Listing 17-4:   Accounting for User Rotation Changes

```
- (void)didRotateFromInterfaceOrientation:
        (UIInterfaceOrientation)fromInterfaceOrientation {

  if (![self.navigationItem.rightBarButtonItem.title
                              isEqualToString: @"Map"])
    [mapView setRegion:
    [self regionForAnnotationGroup:destination.annotations]
                              animated:NO];
}
```

You can see the result of touching the locate button in Figure 17-7. (Ah, Infinite Loop. Apple's Home Sweet Home.)

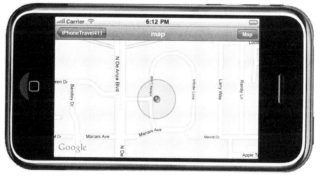

Figure 17-7:
Go to
current
location.

Geocoding

Seeing where I am on the map is all fine and dandy, but stickler that I am, I'd also like to know the exact street address. (If I have the address, I could also write some code to turn the iPhone's current address into an Address Book contact, but I'll allow you the pleasure of figuring that out.)

Being able to go from a coordinate on a map to a street address is called *reverse geocoding,* and thankfully the ability to do that is supplied by the `MapKit`. *Forward geocoding* (also called just geocoding), which converts an address to a coordinate, doesn't come with the `MapKit`, although many free and commercial services that *can* do that are available.

Keep in mind that the location may not be completely accurate — remember that `horizontalAccuracy` business in the "Going to the Current Location" section, earlier in this chapter? For example, because my office is very close to my property line, my location sometimes shows up with my next-door neighbor's address.

Adding reverse geocoding to iPhoneTravel411 will enable you to display the address of the current location. Just follow these steps:

1. **Import the reverse geocoder framework into `MapController.h` and have `MapController` adopt the `MKReverseGeocoderDelegate` protocol.**

   ```
   #import <MapKit/MKReverseGeocoder.h>

   @interface MapController : UIViewController
           <MKMapViewDelegate, MKReverseGeocoderDelegate> {
   ```

2. **Add an instance variable to hold a reference to the geocoder object.**

   ```
   MKReverseGeocoder *reverseGeocoder;
   ```

   You'll use this later to release the `MKReverseGeocoder` after you get the current address.

3. **Add the methods `reverseGeocoder:didFindPlacemark:` and `reverseGeocoder:didFailWithError:` to `MapController.m`.**

   ```
   - (void)reverseGeocoder:(MKReverseGeocoder *) geocoder
           didFindPlacemark:(MKPlacemark *) placemark {

     NSMutableString* addressString;
     if ([selectedAnnotation
                 isKindOfClass:[MKUserLocation class]]){
       if (placemark.subThoroughfare!= NULL) {
         addressString = [[NSMutableString alloc]
             initWithString: placemark.subThoroughfare];
         [addressString appendString: @" "];
         [addressString appendString:
                                 placemark.thoroughfare];
         mapView.userLocation.subtitle =
                                 placemark.locality;
       }
       else {
         addressString = [[NSMutableString alloc]
                 initWithString: placemark.locality];
         mapView.userLocation.subtitle =
                         placemark.administrativeArea;
       }
       selectedAnnotation.title = addressString;
   ```

```
        [addressString release];
    }
}


- (void)reverseGeocoder:(MKReverseGeocoder *) geocoder
        didFailWithError:(NSError *) error{

    NSLog(@"Reverse Geocoder Errored");
}
```

The `reverseGeocoder:didFindPlacemark:` message to the delegate
is sent when the `MKReverseGeocoder` object successfully obtains
placemark information for its coordinate. An `MKPlacemark` object
stores placemark data for a given latitude and longitude. Placemark data
includes the properties that hold the country, state, city, and street
address (and other information) associated with the specified coordinate,
for example. (Several other pieces of data are available that you might
also want to examine.)

- `country`: Name of country

- `administrativeArea`: State

- `locality`: City

- `thoroughfare`: Street address

- `subThoroughfare`: Additional street-level information, such as
  the street number

- `postalCode`: Postal code

In this implementation, you are setting the user location annotation
(`userLocation`) title (supplied by `MapKit`) to a string you create,
made up of the `subThoroughfare` and `thoroughfare` (the street
address). You assign the subtitle the `locality` (city) `property`.

You notice that I do engage in some error checking here. If the

```
placemark.subThoroughfare != NULL)
```

I assume that there's no street address, and instead I set the title to the
city and the subtitle to the state.

Also notice that I check to see whether the annotation is a
`MKUserLocation` object.

```
if ([selectedAnnotation
            isKindOfClass:[MKUserLocation class]]){
```

I determine that by checking the `selectedAnnotation` instance
variable, which I have you define and set in this section.

You may be wondering about why I'm spending so much time on geocoding because, as of now, the only geocoding I do is for the user location. But as you soon see, I also geocode the new location if the user moves the pin on the map.

A placemark is also an annotation and conforms to the MKAnnotation protocol, whose properties and methods include the placemark coordinate and other information. Because they are annotations, you can add them directly to the map view.

The reverseGeocoder:didFailWithError: message is sent to the delegate if the MKReverseGeocoder couldn't get the placemark information for the coordinate you supplied to it. (This is a required MKReverseGeocoderDelegate method.)

Of course, in order to get the reverse geocoder information, you need to create an MKReverseGeocoder object. Make the MapController a delegate, send it a start message, and then release it when you're done with it.

1. **Make the MapController an MKReverseGeocoder delegate by adding the code in bold to MapController.h.**

```
@interface MapController : UIViewController
        <MKMapViewDelegate, MKReverseGeocoderDelegate> {
```

2. **Allocate and start the reverse geocoder and add the MapController as its delegate in the MapController's goToLocation: method by adding the code in bold.**

```
- (IBAction)goToLocation:(id) sender{

  MKUserLocation *annotation = mapView.userLocation;
  CLLocation *location = annotation.location;
  if (nil == location)
    return;
  CLLocationDistance distance =
            MAX(4*location.horizontalAccuracy, 500);
  MKCoordinateRegion region =
            MKCoordinateRegionMakeWithDistance
            (location.coordinate, distance, distance);
  [mapView setRegion:region animated:NO];
  self.navigationItem.rightBarButtonItem.action =
                              @selector(goToTrip:);
  self.navigationItem.rightBarButtonItem.title =
                              @"Map";

  selectedAnnotation = mapView.userLocation;
  reverseGeocoder = [[MKReverseGeocoder alloc]
            initWithCoordinate:location.coordinate];
  reverseGeocoder.delegate = self;
  [reverseGeocoder start];
}
```

Notice how you initialize the `MKReverseGeocoder` with the coordinate of the current location. You also need to add the new instance variable `MKUserLocation   *selectedAnnotation` that I use earlier in the geocoder and assign here.

```
selectedAnnotation = mapView.userLocation;
```

3. **Release the `MKReverseGeocoder` by adding the code in bold to `goToTrip:`.**

```
- (IBAction)goToTrip:(id) sender{

  [reverseGeocoder release];
  [mapView setRegion:[self regionForAnnotationGroup:
        destination.annotations] animated:NO];
  self.navigationItem.rightBarButtonItem.title =
                                    @"Locate";
  self.navigationItem.rightBarButtonItem.action =
                            @selector(goToLocation:);
}
```

You release the `MKReverseGeocoder` in this method because although you start the `MKReverseGeocoder` in the `goToLocation:` method, it actually doesn't return the information in that method. It operates asynchronously; when it either constructs the placemark or gives up, it sends the message `reverseGeocoder:didFindPlacemark:` or `reverse Geocoder:didFailWithError:`, respectively. If you're returning to the original map view, however, you no longer care whether it succeeds or fails because you no longer need the placemark, and you release the `MKReverseGeocoder`.

Figure 17-8 shows the result of your adventures in reverse geocoding.

Figure 17-8:
Reverse
geocoding

But What If I Don't Want to Go to London?

As you are aware, all the destination coordinates and locations are hard-coded in iPhoneTravel411. But in your own apps, you probably would want to give the user control over what's displayed.

In this app, one could argue that it's not really a problem; it's a travel guide to London, after all. But in reality, London is a big place and the user might want to be able to adjust the destination more precisely.

Although you could create a modal dialog to allow the user to add or change a map location, that's beyond the scope of this book. But one thing I show you is how to create a draggable annotation — in this case, the City annotation — so that the user can move his or her destination to exactly where he or she wants to go.

The way to do that is pretty simple. All you do is set an annotation view property. Of course, to do that you're going to need access to the annotation view. So, instead of using the default view, as you have been doing, I show you how to use an SDK-supplied annotation view — MKPinAnnotationView. In addition, you'll also be able to set the pin color as well as animate the pin to drop on to the map.

All you have to do is add the code in Listing 17-5 to MapController.m. mapView:viewForAnnotation: is a MKMapView delegate method that is automatically invoked before the map view displays the annotation and gives you a chance to customize the view accordingly.

Listing 17-5:   mapView:viewForAnnotation:

```
- (MKAnnotationView *)mapView:(MKMapView *)aMapView
        viewForAnnotation:(id <MKAnnotation>)annotation {

    if ([annotation isKindOfClass:[MKUserLocation class]])
        return nil;
    MKPinAnnotationView* pinView = (MKPinAnnotationView*)
[mapView dequeueReusableAnnotationViewWithIdentifier:
                        @"CustomPinAnnotationView"];
    if (!pinView) {
        pinView = [[[MKPinAnnotationView alloc]
            initWithAnnotation:annotation
            reuseIdentifier:@"CustomPinAnnotation"]
                                        autorelease];
        if ([annotation isKindOfClass:[City class]]) {
```

```
      pinView.pinColor = MKPinAnnotationColorRed;
      pinView.draggable =YES;
    }
    else
      pinView.pinColor = MKPinAnnotationColorGreen;
    pinView.animatesDrop = YES;
    pinView.canShowCallout = YES;
  }
  else
    pinView.annotation = annotation;
  return pinView;
}
```

I start by checking to see whether the annotation is a MKUserLocation. If it is, I just use the built-in view.

```
if ([annotation isKindOfClass:[MKUserLocation class]])
    return nil;
```

The next thing I do is check to see whether there's a view lying around that I can use. This is the same kind of mechanism you use to reuse table cells in Chapter 15. If there isn't a view available, I create one, initializing it with the annotation.

```
if (!pinView) {
  pinView = [[[MKPinAnnotationView alloc]
    initWithAnnotation:annotation
    reuseIdentifier:@"CustomPinAnnotation"] autorelease];
```

If the annotation is a City object, I set the pin color to red — that is the "customary" color for a destination. I also make the pin draggable. Setting this property to YES (the default value of this property is NO) does what you'd expect it to do — make an annotation draggable by the user — but if you do that, the annotation object must also implement the setCoordinate: method. If you look at the code for the City model on my Web site, www.nealgoldstein.com, you can see that I've defined coordinate as a property and coded the @synthesize statement.

If it's not a City object — it's an airport, for example — I'd set the pin color to green, the customary color for an origin. I don't want that pin to be draggable, so I won't bother there with the draggable property.

Here's the code I came up with:

```
if ([annotation isKindOfClass:[City class]]) {
    pinView.pinColor = MKPinAnnotationColorRed;
    pinView.draggable =YES;
  }
  else
    pinView.pinColor = MKPinAnnotationColorGreen;
```

Finally, I make the pin drop animated and allow it to show a callout, which will be handled by the pin view.

```
pinView.animatesDrop = YES;
pinView.canShowCallout = YES;
```

If there were a reusable pin view, I assign the annotation, and in either case, return the pin view.

```
else
   pinView.annotation = annotation;
   return pinView;
```

Of course, if the user does move the destination, it would be nice to change the title and subtitle to the new location. Fortunately, there's another delegate method that gives me a chance to do just that — mapView:annotationView: didChangeDragState:fromOldState:.

Listing 17-6 shows the code you need to add to mapController.m in order to get this title/subtitle change stuff to work for you.

Listing 17-6:   mapView:annotationView:didChangeDragState:fromOldState:

```
- (void)mapView:(MKMapView *)mapView annotationView:
    (MKAnnotationView *)annotationView
   didChangeDragState:(MKAnnotationViewDragState)newState
       fromOldState:(MKAnnotationViewDragState)oldState {

   if (newState == MKAnnotationViewDragStateEnding) {
     reverseGeocoder = [[MKReverseGeocoder alloc]
            initWithCoordinate:
                    annotationView.annotation.coordinate];
     selectedAnnotation = annotationView.annotation;
     reverseGeocoder.delegate = self;
     [reverseGeocoder start];
   }
}
```

The mapView:annotationView:didChangeDragState:fromOldState: message is constantly sent to the delegate as the user drags the annotation hither and yon. Although you're constantly being updated, I really don't care where the user has dragged it until the dragging is over.

```
if (newState == MKAnnotationViewDragStateEnding) {
```

When the dragging is over, though, I save the annotation in selected Annotation and then create a geocoder to go with the new location

```
selectedAnnotation = annotationView.annotation;
reverseGeocoder.delegate = self;
[reverseGeocoder start];
```

As you recall, in the `reverseGeocoder:didFindPlacemark:` method, you updated the `MKUserLocation` annotation title and subtitle with the location placemark information. All I have to do now is add the code in bold in Listing 17-7 to update a `City` annotation.

Listing 17-7:   Adding to reverseGeocoder:didFindPlacemark:

```
- (void)reverseGeocoder:(MKReverseGeocoder *) geocoder
          didFindPlacemark:(MKPlacemark *) placemark {
  NSMutableString* addressString;
  if ([selectedAnnotation
              isKindOfClass:[MKUserLocation class]]){
    if (placemark.subThoroughfare!= NULL) {
      addressString = [[NSMutableString alloc]
          initWithString: placemark.subThoroughfare];
      [addressString appendString: @" "];
      [addressString appendString:
                            placemark.thoroughfare];
      mapView.userLocation.subtitle =
                            placemark.locality;
    }
    else {
      addressString = [[NSMutableString alloc]
                  initWithString: placemark.locality];
      mapView.userLocation.subtitle =
                        placemark.administrativeArea;
    }
    selectedAnnotation.title = addressString;
  }
  else {
    if (placemark.locality!= NULL)
      addressString = [[NSMutableString alloc]
                  initWithString: placemark.locality];
    else
      addressString = [[NSMutableString alloc]
        initWithString: placemark.administrativeArea];
    [(City*)selectedAnnotation
                    setCityName:addressString];
  }
  [addressString release];
//  }
}
```

To update the city name, I send the set `setCityName:` message, which was generated by the `@synthesize` statement you entered when you defined the property.

Here of course I could have used the dot notation, but using the method makes the casting less cumbersome.

As you can see in Figure 17-9, I've dragged the pin to Kensington, where I'm planning to stay with my friend

Figure 17-9:
Staying with
friends.

What's Next?

At this point, you have a fairly robust application. But never being accused of willing to leave well enough alone, I want to show you even more location-based features — which is why I've added a whole new Chapter 18.

Chapter 18

Location, Location, Location

*W*hen I land at Heathrow, I just know I'm going to be tired and cranky. All I'll want to do is get to my hotel, take a shower, and climb into bed for a short nap. One of the things that will keep me going is knowing how far I am from my hotel. In fact, I wouldn't mind watching the miles count down as I get closer. Because I'm going to take the Paddington express, I'll also have to take a cab or the tube to my hotel. If I take a cab, it would be nice to be able to watch the distance to my hotel decrease over time and not be taken on a tour of London.

And one other thing: I'd also like to be able to have iPhoneTrvel411 keep me updated with local notifications whilst it runs in the background, so I can be doing other things on my iPhone as I sit there (like play that clever game I wrote before I left, or check my e-mail).

All this is indeed possible in iOS 4. In this chapter, I explain how to use background processing, core location, and local notifications to do all those things. Using iOS 4, you have several alternatives when it comes to getting your location — alternatives that trade off accuracy for battery life — so I fill you in on your options. I even show you a feature available in iPhone 4 that limits your notifications to a particular region.

Doing all this is relatively simple, but the rules you have to follow in order to have your app run in the background — as well as the number of options you have — means that you have to pay attention to the details.

I want to start with core location and then I get to how you can get your app to run in the background.

Being Location Aware

There is a class of applications (like travel and social networking apps) where knowing the user's geographic location can help you improve the quality of the information you offer. And in fact, mobile devices such as the iPhone are about mobility (duh) and the fact that your application is running on a device that can go anywhere. Location services supports that by providing the information about where the device is.

In Chapter 17, I explain that there are two parts to location services: mapping (explained in Chapter 17) and core location (soon to be explained in this very chapter).

But first a little background.

The iPhone has three positioning technologies it can use to determine its location:

1. Cell positioning
2. Wi-Fi positioning
3. GPS positioning

These are in order of both accuracy and power consumption, with Cell positioning being the lowest and GPS positioning the highest. They are all complementary.

Cell positioning can locate the user even without a connection and is accurate to 0.5 to 50 kilometers. The more cells, the more accurate — so it works best in urban areas. It works for all devices with a cell radio.

Wi-Fi works for all devices. It's based on the location of Wi-Fi hotspots; there are also caching techniques that are used to even allow positioning when an Internet connection is not available. The accuracy is often in the 100-meter range.

Finally there's GPS, based on GPS satellites. It also includes some assisting technology that can increase accuracy. Accuracy depends on a number of factors, but you can figure around 8 meters.

Even if the user has roaming turned off, he or she can still get the benefit of location services.

Regardless of which technology is being used, though, you have to deal with only one API — the classes of the Core Location framework. That means you generally don't have to be concerned about which technology(ies) the device is using.

Core Location provides three distinct services you can use to get and monitor the device's current location:

- ✔ **The standard location service:** This is very configurable and is the most accurate and uses the most power. It's supported in all versions of iPhone OS.

- ✔ **The significant-change location service:** This service was added with iOS 4. It uses cell tower positioning to notify you of significant changes in location (I explain what that means later). It's less accurate, but also uses less power than the standard location service.

- ✔ **Region monitoring:** On the iPhone 4 (and later), this service lets you monitor when the user enters or leaves a region that you define.

To use the features of the Core Location framework, you have to add `CoreLocation.framework` to your Xcode project (just as you did with `MapKit,framework`), as well as add a `#import <CoreLocation/CoreLocation.h>` statement to all the files that use the Core Location classes. Your iPhoneTravel411AppDelegate will also need to become a delegate by using the `CLLocationManagerDelegate` protocol.

I have you do all your location work in `iPhoneTravel411AppDelegate`, so add the statements in bold to `iPhoneTravel411AppDelegate.h`:

```
#import <CoreLocation/CoreLocation.h>

@interface iPhoneTravel411AppDelegate : NSObject
      <UIApplicationDelegate, CLLocationManagerDelegate> {
```

You're doing all of this in `iPhoneTravel411AppDelegate` so that you can create a single set of services within your application that supports Core Location. Admittedly, instead of doing all of this in `iPhoneTravel411 AppDelegate`, you could just create a new class that does the same things and also encapsulates the functionality. Because I use Core Location extensively in my own apps, I created a class that I use in my own apps in the store.

Gathering location data is a power-intensive operation. How intensive? Well, I cover that in the "Using the Significant-Change Location Service to Reduce Battery Drain" section, later in this chapter.

Getting the user's current location by using the standard location service

To start, you first need to determine whether location services are available. Although every iOS device can support location services, there are still going to be situations where the service is not available. For example,

✔ The user can disable location services in the Settings application.

✔ The user can deny location services for a specific application.

✔ The device might be in Airplane mode.

✔ On devices without GPS, there may be no cell towers or Wi-Fi hotspots in the area.

You need to code for these possibilities in your application. I explain some things to get you started, but to create a seamless user experience, you have to do more work on your own.

Time for the overview: To add location services to iPhoneTravel411, you have to do three things:

1. Start the location manager.

2. Respond to changes in location.

3. Handle errors.

The next few sections give the details.

Starting the location manager

Start by firing up the location service. You need to do that in two places — `application:didFinishLaunchingWithOptions` and `setDefaults:`. To get this bit of business started, add the code in bold in Listings 18-1 and 18-2 to those two methods in `iPhoneTravel411AppDelegate.m`.

Listing 18-1:   Starting the Location Manager in the App Delegate

```
- (BOOL)application:(UIApplication *)
   application didFinishLaunchingWithOptions:
                          (NSDictionary *)launchOptions {

   [[NSNotificationCenter defaultCenter] addObserver:self
      selector:@selector(setDefaults:)
      name:NSUserDefaultsDidChangeNotification object:nil];

   if (![[NSUserDefaults standardUserDefaults]
         objectForKey:kUseStoredDataPreference]) {
      useStoredData = NO;
      monitorLocationChanges = YES;
      [self startLocationUpdates:monitorLocationChanges];
   }
   else
      [self setDefaults:nil];
   ...
}
```

Now for `setDefaults`. As you may recall from Chapter 16, when your application is launched, you check to see whether preferences are available. If not, you send yourself the `startLocationUpdates:` messages with an argument of `YES` (the default).

Listing 18-2: Responding to Preference Change

```
- (void)setDefaults:(NSNotification*)notification {

    useStoredData = [[NSUserDefaults standardUserDefaults]
            boolForKey:kUseStoredDataPreference];
    monitorLocationChanges = [[NSUserDefaults
            standardUserDefaults] boolForKey:kMonitorLocati
            onPreference];
    [self startLocationUpdates:monitorLocationChanges];
}
```

In this case, you've received a notification that the preference has changed (or your application was launched and preferences are available), and you pass that state to `startLocationUpdates:`. You should also do that for stored data, although you'll have to set the stored controllers to `nil` in the menu list.

Because `startLocationUpdates:` appears to be where the action is, let's look at it. After you add the Core Location framework and the import statements, as spelled out at the beginning of this chapter, add the code in Listing 18-3 to `iPhoneTravel411AppDelegate.h`.

Listing 18-3: Starting the Standard Service

```
- (void)startLocationUpdates:(BOOL) startUpdates {

    if (startUpdates) {
      if (nil == locationManager)
        locationManager = [[CLLocationManager alloc] init];
        locationManager.delegate = self;
        locationManager.purpose = @"iPhoneTravel411 will keep
            track of how far you are from your destination.";
        locationManager.desiredAccuracy =
                                    kCLLocationAccuracyKilometer;
        locationManager.distanceFilter = 500;
        [locationManager startUpdatingLocation];
    }
    else {
      [locationManager stopUpdatingLocation];
    }
}
```

Walking through the code, you can see that the first thing I do is check to see whether I should start or stop the location manager.

```
if (startUpdates) {
```

One thing I don't do here is check to see whether location services are available. I do that by sending the `locationServicesEnabled` message to the `CLLocationManager` class. (In iPhone OS 3.x and earlier, check the value of the `locationServicesEnabled` property instead.) If location services aren't available, and you start location services anyway, the system will ask the user whether location services should be reenabled. On the one hand, this can be annoying to the user if he or she has location services (still) disabled intentionally. However, if your application requires that services be enabled, this makes it easier for the user to reenable them.

Then I check to see whether I already have an instance. If not, I go ahead and create one.

```
if (nil == locationManager)
    locationManager = [[CLLocationManager alloc] init];
```

Next, I assign myself as the delegate. The delegate will be receiving the location updates — I show you that next.

```
locationManager.delegate = self;
```

To be polite, I tell the user why I want to track location. This is added to the alert that asks the user whether it's okay for an app to track his or her location. You can see that message in Figure 18-1. (You only see it on the device).

```
locationManager.purpose = @"iPhoneTravel411 will keep
        track of how far you are from your destination.";
```

The next thing you do is set the desired accuracy and a distance filter:

```
locationManager.desiredAccuracy =
                        kCLLocationAccuracyKilometer;
locationManager.distanceFilter = 500;
```

You can use a number of constants here for accuracy. The most accurate one is `kCLLocationAccuracyBestForNavigation`, which also uses additional sensor data. However, this level of accuracy is really only intended for applications that require precise position information at all times and should be used only while the device is plugged in. If that's not the case, you should use `kCLLocationAccuracyBest`.

Although the receiver does its best to achieve the requested accuracy, it's not guaranteed.

Use what's appropriate for your app's needs here. The more accuracy you need, the more power is required and the shorter the battery life.

Even when you do specify high-accuracy data, the initial event delivered by the location service may not have the accuracy you requested. Location Services tries to deliver the initial event as quickly as possible and then continues working and delivers additional events, as better data becomes available.

The distance filter specifies minimum distance (measured in meters) a device must move laterally before an update event is generated, measured relative to the previously delivered location. Be careful with this value as well. If you play around with this value in your app and set it too low, you quickly see how annoying frequent updates can be.

If you do want to know every time the user takes a step, use kCLDistance FilterNone. This is also the default.

Both of these properties are used only by the standard location services and are not used with the significant-change location service (which I explain shortly).

Finally, you tell the location manger to start up:

```
[locationManager startUpdatingLocation];
```

If the value passed in was NO, you turn off the location manger:

```
[locationManager stopUpdatingLocation];
```

You also need to add the following two lines of code to iPhoneTravel 411AppDelegate.h to declare the instance variable and a method you will be using. The first should be within the braces along with other instance variables. The second should be with the methods and property declarations between the closing brace and @end

```
    CLLocationManager        *locationManager;
- (void)startLocationUpdates: (BOOL) startUpdates;
```

Figure 18-1:
Why you
want to
track the
user.

Receiving location data from a service

The location manger sends messages to your delegate when it receives updates. That's what it does. Now, whether you use the standard location service (or the significant-change location service I show you next) to get location events, the way you receive those events is the same. Whenever a new event is available, the location manager sends the location Manager:didUpdateToLocation:fromLocation: message to its delegate. If there's an error retrieving an event, the location manager sends the locationManager:didFailWithError: message to its delegate instead.

You can also get heading information in a device that includes a compass by sending the startUpdatingHeading message to the location manager. Heading events are delivered to the locationManager:didUpdateHeading: method of your delegate. If there's an error, the location manager calls the locationManager:didFailWithError: method of your delegate instead.

Listing 18-4 shows the `locationManager:didUpdateToLocation:from Location:` delegate method. Apple suggests that, because the location manager object sometimes returns cached events, you'd be smart to check the timestamp of any location events you receive. In this code, you do that and then throw away any events that are more than 15 seconds old.

Listing 18-4:   Processing an Incoming Location Event

```
- (void)locationManager:(CLLocationManager *)manager
        didUpdateToLocation:(CLLocation *)newLocation
        fromLocation:(CLLocation *)oldLocation {

  NSDate* eventDate = newLocation.timestamp;
  NSTimeInterval howRecent = [eventDate
        timeIntervalSinceNow];
  if (abs(howRecent) < 15.0)
  {
    NSLog(@"latitude %+.6f, longitude %+.6f\n",
        newLocation.coordinate.latitude,
        newLocation.coordinate.longitude);
  }
}
```

To ascertain the time at which this location was delivered, you access the `timestamp` property of the `CLLocation` object.

```
  NSDate* eventDate = newLocation.timestamp;
```

To find out how long ago this event was (in seconds), you send the `time IntervalSinceNow` message to the `NSDate` object.

```
  howRecent = [eventDate timeIntervalSinceNow];
```

I haven't said anything about the `NSDate` class, which provides date and time information for you. It's beyond the scope of this book, but it's something you should explore on your own.

After you add the code and compile it, you get all sorts of nice messages on the log, like this one:

```
  2010-06-27 07:36:37.156 iPhoneTravel411[94443:207]
                  latitude +37.331650, longitude -122.030704
```

Of course, printing out the location on the log doesn't help the user much. You'll want to send him or her a notification, and I show you that in the "Local Notifications" sections soon. But first . . .

Dealing with failure

If there's an error retrieving an event, the location manager will send the locationManager:didFailWithError: message to its delegate.

To take care of that scenario, add the code in Listing 18-5 to iPhoneTravel411AppDelegate.m.

Listing 18-5: Failure

```
- (void)locationManager:(CLLocationManager *)manager
    didFailWithError:(NSError *)error {

        NSLog(@"Location error %@, %@", error, @"Fill in
        the reason here");
}
```

All I do is print the error message to the log, but you may want to do something else here (like tell the user all about it).

error is an object containing the reason why the location or heading couldn't be retrieved and has an error code that tells you what the reason is. (You'll find that most of the system services classes handle errors in this way.)

Some of the possible errors are

✔ kCLErrorLocationUnknown when the location service can't get a location fix right away

✔ kCLErrorDenied when the user denies your application's use of the location service

Local Notifications

Okay, now that you have your location information, what do you want to do with it?

Of course, although you as a developer can while away the hours watching the location changes by using an NSLog (don't laugh — I actually kept my iPhone attached to my MacBook Pro and drove around to make sure that things worked), your user needs a more user-friendly way of getting updated. You can offer that easy updating to the user through local notifications.

Local notifications work (that is, you can post and receive them) not only when the application is running in the foreground, but also when it's running in the background. So, I need to explain how this works in both circumstances. I start with the Foreground scenario and in the next section reveal how multitasking in iOS 4 enables your application to process location events and post notifications while running in the background.

Listing 18-6 starts things off by showing you how to post a local notification.

Listing 18-6: Creating the Notification

```
- (void)locationChangeNotify {

  UILocalNotification *note
                   = [[UILocalNotification alloc] init];
  RootViewController *theRootViewController =
    [navigationController.viewControllers objectAtIndex:0];
  CLLocationCoordinate2D cityCoordinate =
    [theRootViewController.destination returnCityLocation];
  CLLocation *cityLocation =
  [[CLLocation alloc] initWithLatitude:
                       cityCoordinate.latitude
                       longitude:cityCoordinate.longitude];
  NSString *message = [[NSString alloc]initWithFormat:
    @" You are now %.1f miles from %@",
    [locationManager.location
    distanceFromLocation:cityLocation]/kMetersToMiles,
    [theRootViewController.destination returnCityName]];
  note.alertBody= message;
  [message release];
  [[UIApplication sharedApplication]
                   presentLocalNotificationNow:note];
  [note release];
  [cityLocation release];
}
```

Note that I start here by allocating and initializing a `UILocalNotification` object.

```
UILocalNotification *note
                   = [[UILocalNotification alloc] init];
```

You can set a number of properties in the notification, including the `fire Date` — the date and time that the OS should deliver the notification. If the value is the `fireDate` is `nil`, (which is the default) the notification is delivered immediately.

You can also set an icon badge number and sound, as well as attach custom data to the notification through the userInfo property. I won't do any of that here, but such bells and whistles are definitely an option.

Continuing through Listing 18-6, you see that I get the location of the City object (this started as London, but you made it draggable in Chapter 17) by getting the destination object from the RootViewController (the RootViewController.destination) and sending it the returnCity Location message. That kind of thing should be old hat to you by now.

What's interesting is how I find the RootViewController. The iPhone Travel411AppDelegate doesn't have an instance variable pointing to the RootViewController. While your first thought may be to try and figure out how to create one, there is an easier way. What iPhoneTravel411AppDelegate *does* have is an instance variable pointing to the NavigationController. As I explain in Chapter 14, the navigation controller maintains a stack of view controllers, one for each of the views displayed, starting with the root view controller, and — voilà!

I then create a CLLocation object, which I use shortly, and assign it to cityCoordinate.

```
RootViewController *theRootViewController =
    [navigationController.viewControllers objectAtIndex:0];
CLLocationCoordinate2D cityCoordinate =
    [theRootViewController.destination returnCityLocation];
CLLocation *cityLocation =
  [[CLLocation alloc] initWithLatitude:
          cityCoordinate.latitude
          longitude:cityCoordinate.longitude];
```

I then create the message note.alertBody, specifying how far I am from the city, by sending distanceFromLocation message to the location Manager.location object (which has the current location) and using the CLLocation object cityLocation (which I just created) as the argument.

distanceFromLocation: is one of my favorite methods, because it returns the distance in meters between any two CLLocation objects (this is why I had to create a CLLocation object by using the CLLocationCoordinate2D returned by the City object). This saves a lot of trigonometry (which I actually used to do before I found this method). I also add a new constant to Constants.h (#define kMetersToMiles 1609.344) so that I can convert the meters to miles.

```
NSString *message = [[NSString alloc]initWithFormat:
    @" You are now %.1f miles from %@",
    [locationManager.location
    distanceFromLocation:cityLocation]/kMetersToMiles,
    [[theRootViewController destination] returnCityName]];
note.alertBody= message;
[message release];
```

I then tell the OS to post the local notification by sending the UIApplication message scheduleLocalNotification:. As I say earlier in this section, the application uses the fireDate property of the UILocalNotification object to know when to deliver it. Because I left that as nil, it's delivered immediately. I could also have sent the presentLocalNotificationNow: message, which would have sent the notification immediately.

To dot your *i*'s and cross your *t*'s, you also need to add the declaration in iPhoneTravel411AppDelegate.h.

Of course, to actually see the local notification, you do have to send yourself the locationChangeNotify message and then do something with it. You send yourself the message when you get a location update in location Manager:didUpdateToLocation:fromLocation: you just added. To set that in motion, add the code in bold in Listing 18-7.

Listing 18-7: Requesting a Notification

```
- (void)locationManager:(CLLocationManager *)manager
        didUpdateToLocation:(CLLocation *)newLocation
        fromLocation:(CLLocation *)oldLocation {

  NSDate* eventDate = newLocation.timestamp;
  NSTimeInterval howRecent = [eventDate
                                    timeIntervalSinceNow];
  if (abs(howRecent) < 15.0) {
    NSLog(@"latitude %+.6f, longitude %+.6f\n",
        newLocation.coordinate.latitude,
        newLocation.coordinate.longitude);
  }
  [self locationChangeNotify];
}
```

The final piece is receiving the notification and doing something with it to inform the user of what is going on. The "doing something with it" part is taken care of by another delegate method, application:didReceiveLocal Notification:. Listing 18-8 shows the details.

Listing 18-8: Processing the Notification When in the Foreground

```
- (void)application:(UIApplication *)application
        didReceiveLocalNotification:
                        (UILocalNotification *)notification {

  if ([UIApplication sharedApplication].applicationState
                        == UIApplicationStateActive) {
    UIAlertView *alertView = [[UIAlertView alloc]
      initWithTitle:@"You are on your way"
      message:notification.alertBody delegate:nil
      cancelButtonTitle:@"OK" otherButtonTitles:nil];
    [alertView show];
    [alertView release];
  }
}
```

Walking through Listing 18-8, you see that when the application is running in the foreground, the app delegate is sent the `application:didReceiveLocal Notification:` message (it is sent — no surprise here — the `application: didReceiveRemoteNotification` message for remote notifications) and passes in the local notification object.

In `application:didReceiveLocalNotification:`, you can do all sorts of things that are beyond the scope of this book, such as reset a badge number and so on.

In this case, all I do is check to see whether I'm active — meaning the application is running in the foreground — and if I am, I post an alert:

```
if ([UIApplication sharedApplication].applicationState
                        == UIApplicationStateActive) {
  UIAlertView *alertView = [[UIAlertView alloc]
    initWithTitle:@"You are on your way"
    message:notification.alertBody delegate:nil
    cancelButtonTitle:@"OK" otherButtonTitles:nil];
  [alertView show];
  [alertView release];
}
```

Keep in mind that you could just as easily be running in the background, and if you were, you wouldn't want to create an alert or any other user-interface object.

When all is said and done, you'll see more or less what is shown in Figure 18-2. I say more or less because, as you've certainly noticed, I moved the destination to Harrow on the map, and that is what's displayed on the alert.

If you aren't active, you allow the OS to post the alert you see in Figure 18-3. Being able to post an alert while not the foreground application comes to you courtesy of multitasking, and I cover that next.

Finally, you also have to add `returnCityName` and `returnCityLocation` to `Destination.m` and their declarations in `Destination.h` and `#import Destination.h` into `iPhoneTravel411AppDelegate.m`

```
- (CLLocationCoordinate2D) returnCityLocation {

    return city.coordinate;
}

- (NSString*) returnCityName {

    return city.cityName;
}
```

Figure 18-2:
How far
from
wherever.

Figure 18-3:
A system
alert.

Running in the Background

Of course, even though I would love to be reminded of my progress to my destination, I don't want to have to have iPhoneTrvael411 always in the foreground in order to do that. One of the great things about the multitasking available in iOS 4 on the iPhone 3GS and above is that you can now process certain kinds of events in the background. And what's even better is that location updates are one of those kinds of events.

If you want the application to have location updates delivered in the background, you have several choices:

✔ You can continue to use the standard location services, which I cover in this section.

or

✔ You can also choose to use the significant-location change service, which will wake your application at appropriate times to handle new events.

I explain more about why you may want to do this — and how to do it — in the "Using the Significant-Change Location Service to Reduce Battery Drain" section, later in this chapter.

An application should request background location services only if it's really necessary, and it's up to you to decide whether you need to do that in order to deliver the best possible user experience.

In Chapter 2, I explain how (certain) applications can be moved into the background if the user does certain things:

✔ The application is interrupted by an incoming phone call, SMS, or calendar notification.

✔ The user presses the Sleep/Wake button

✔ The user presses the Home button, or the system launches another application, and the application is moved to background.

Applications built with iPhone SDK 4.0 or later (and running in iOS 4.0 and later) are no longer terminated when these things happen. Instead, they now shift to a background execution context. Although most applications are suspended shortly after moving to the background, applications that need to continue working in the background can do one of the following to continue to execute:

1. Request a finite amount of time to complete a task.

2. Declare that it supports specific services that require background execution.

3. Use local notifications to generate user alerts at designated times, whether or not the application is running.

I cover Option 2 in this section and Option 3 in the "Using the Significant-Change Location Service to Reduce Battery Drain" section, later in this chapter.

Background applications can still be terminated under certain conditions (such as during low-memory conditions), and so applications must be ready to exit at any time. This means that many of the things you used to do at termination must be done when your application moves to the background.

The ability to run background tasks is not supported on all iOS-based devices. If a device isn't running iOS 4 and later, or if the hardware can't support multitasking, the system handles application termination differently. If you want to know how to create an application that can run under pre–iOS 4 OSes, check out my Web site at www.nealgoldstein.com.

Declaring the Background Tasks You Support

If you want to execute in the background, you have to *tell* the OS you want to do it. You do that by including the UIBackgroundModes key in its Info. plist file. This key identifies which background tasks your application supports and can have the following values:

- ✔ **Audio:** The application plays audio in the background

- ✔ **Location:** The application processes location events in the background.

- ✔ **VoIP:** The application provides the ability for the user to make Voice over IP calls.

So how do you do set it up so that iPhoneTravel411 can support location services in the background?

In the Groups & Files list of Xcode's Project window, select iPhoneTravel 411-info.plist and then select the last entry in the Editor view, as I have in Figure 18-4.

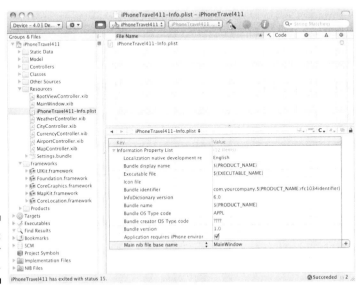

Figure 18-4: Info.plist for your app.

This is similar to the root.plist you worked with in Chapter 15 to add a preference.

Select the + icon to the right and then scroll down to the Required Background Modes entry. Expand the disclosure triangle in front of Required Background Modes and choose App Registers for Location Updates from the drop-down menu. You can see the final result in Figure 18-5.

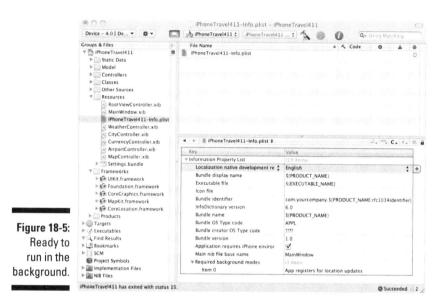

Figure 18-5:
Ready to run in the background.

That's all you have to do; from now on, you'll continue to get location events even when you're in the background.

As I mention earlier, though, there's an alternative to running and constantly receiving alerts. You can also use the significant-change location service, which means that even if you're not receiving location updates in the background, your application will be activated (and even launched, as necessary) to process a perceived change.

Using the Significant-Change Location Service to Reduce Battery Drain

As you may have figured out based on my harping about the battery, getting location information uses a lot of power, and having the standard location service running can drain the device's battery. Most applications don't need location services to be running all the time, so turning off those services is the simplest way to save power. And you can always start location services again later if needed.

You should also configure the parameters of the standard location service in a way that minimizes its impact on battery life. Use the lowest possible setting for desired accuracy — remember that the higher the accuracy, the more power is used. Unless your application really needs to know the user's position within a few meters, do not put the values `kCLLocation AccuracyBest` or `kCLLocationAccuracyNearestTenMeters` in the `desiredAccuracy` property. And remember that specifying a value of `kCLLocationAccuracyThreeKilometers` doesn't prevent the location service from returning better data. Most of the time, Core Location can return location data with an accuracy within 100 meters or so by using Wi-Fi and cellular signals.

You can also use the significant-change location service, as spelled out in the next section. It uses much less power because it monitors only cell tower changes, but you do need a cell radio for it to work. When you can use it, the significant-change location service provides significant power savings while still allowing you to leave location services running. You want to do that in applications that track changes in the user's location but don't need the higher precision offered by the standard location services. And, as you'll see, the significant-change location service can wake your suspended application, or even relaunch it, when a location event comes in — you don't necessarily have to run the app in the background.

In addition to all the things you as a developer can do to reduce battery usage, you should also give the user control as well. This is the reason I had you add the user preference, Monitor Distance, in Chapter 15. This allows the user to decide how useful this feature is and to turn it off if he or she is concerned about battery life.

Using the Significant-Change Service

Using the significant-change location service is almost exactly the same as using the standard location service. Changes are sent to the same delegate method `locationManager:didUpdateToLocation:fromLocation:` — but this time only significant changes registered. Similarly, errors are sent to the very same `locationManager:didFailWithError:` method. In this application, if you wanted to use the significant-change location service, the code in those two methods would stay the same. The only difference would be that instead of sending the location manager the `startUpdating Location` and `stopUpdatingLocation` messages, you would send the `startMonitoringSignificantLocationChanges` and `stopMonitoring`

`SignificantLocationChanges`, as you can see from the code in bold in Listing 18-9. If you wanted to use the significant-change location service, you would delete the code with the strikethrough and add the code in bold.

Listing 18-9: The Significant-Change Location Service

```
- (void)startLocationUpdates:(BOOL) startUpdates {
  if (startUpdates ) {
    if (nil == locationManager)
      locationManager = [[CLLocationManager alloc] init];
    locationManager.delegate = self;
    locationManager.purpose = @"iPhoneTravel411 will
          keep track of how far you are from your
          destination.";
    locationManager.desiredAccuracy =
          kCLLocationAccuracyKilometer;
    locationManager.distanceFilter = 500;
    [locationManager startUpdatingLocation];
    [locationManager
              startMonitoringSignificantLocationChanges];
  }
  else {
    [locationManager stopUpdatingLocation];
    locationManager
              stopMonitoringSignificantLocationChanges];
  NSLog(@"Monitoring stopped");
  }
}
```

You also wouldn't need to add that business about App Registers for Location Updates in the Required Background Modes entry I mention in the "Declaring the background tasks you support" section, earlier in the chapter, for the simple reason that your application doesn't have to be running in the background. It will be woken and given time in the background to process the `locationManager:didUpdateToLocation:fromLocation:` message. If your application has been terminated, it will be relaunched in the background to process the event.

Region Monitoring

In iPhone 4 and later, applications can use region monitoring to be notified when the user crosses geographic boundaries. You can use this capability to generate alerts when the user leaves or enters an area. This means that your app can send people like me, before they get too far from home, one last reminder to check that they have their iPhone, passport, and tickets.

This is something you can actually code on your own — and I've done so in some of my own applications — but using location services to do it for you is much easier.

One note: At this point, with so few iPhone 4 devices out there, it's not a feature you need to worry about in your application, unless you implement a version on your own to use in those devices when the hardware isn't available.

The implementation here is actually very similar to the significant-change location service. The region associated with your application is tracked at all times, including when your application isn't running. If a region boundary is crossed while an application isn't running, that application is relaunched into the background to handle the event. (I show you that later.) Similarly, if the application is suspended when the event occurs, it's woken up and given a short amount of time to handle the event.

Listing 18-10 show you how you could implement region monitoring in iPhoneTravel411.

Listing 18-10:   Region Monitoring

```
- (BOOL)startRegionMonitoring {
  if (![CLLocationManager regionMonitoringAvailable] ||
      ![CLLocationManager regionMonitoringEnabled] )
    return NO;

  CLLocationCoordinate2D home;
  home.latitude = +37.441655;
  home.longitude = -122.143282;
  CLRegion* region = [[CLRegion alloc] initCircularRegionW
          ithCenter:home
                    radius:100.0 identifier:@"home"];
  if (locationManager == nil)
    locationManager = [[CLLocationManager alloc] init];
  [locationManager startMonitoringForRegion:region
          desiredAccuracy:kCLLocationAccuracyBest];

  [region release];
  return YES;
}

- (void)locationManager:(CLLocationManager *)manager
        didEnterRegion:(CLRegion *)region {

  [self leavingHomeNotify];
}
```

```
- (void)locationManager:(CLLocationManager *)manager
        didExitRegion:(CLRegion *)region {

  [self leavingHomeNotify];
}

- (void)locationManager:(CLLocationManager *)manager
        monitoringDidFailForRegion:(CLRegion *)
        regionwithError:(NSError *)error {
        NSLog(@"Location error %@, %@", error, @"Fill in
        the reason here");
}

- (void)leavingHomeNotify {

        UILocalNotification *note =
        [[UILocalNotification alloc] init];
        note.alertBody= @"Don't forget your iPhone,
        Tickets, and passport";
  [[UIApplication sharedApplication] presentLocalNotificat
        ionNow:note];
        [note release];
}
```

I'm not going to go over every line of code here, because it really is beyond the scope of this book, but I do want to point out a few things.

Before I start to monitor any regions, I check to see whether region monitoring is even supported on the current device.

```
if (![CLLocationManager regionMonitoringAvailable] ||
    ![CLLocationManager regionMonitoringEnabled] )
  return NO;
```

There are several reasons why region monitoring might not be available:

✔ The device may not have the hardware that supports region monitoring.

✔ The user may have disabled location services.

✔ The device might be in Airplane mode.

It pays, then, to see whether the service is available. If it isn't, you could either not display this as an app function, or you could implement a software version of it in your app by defining your own region and then checking every time there is a location change to see if the user is no longer in that region (and drain more battery power in the process).

If I want to monitor a region, I have to first define it and register it with the system. Regions are defined by using the CLRegion class, which currently supports the creation of circular regions. A region you create must include a definition of the geographic area and a unique identifier string, which you then use to identify a region later. I define the region center with the coordinates of my home, and give it a radius of 100 meters.

```
CLLocationCoordinate2D home;
   home.latitude = +37.441655;
   home.longitude = -122.143282;
   CLRegion* region = [[CLRegion alloc] initCircularRegionW
        ithCenter:home
                     radius:100.0 identifier:@"home"];
   if (locationManager == nil)
     locationManager = [[CLLocationManager alloc] init];
   [locationManager startMonitoringForRegion:region
              desiredAccuracy:kCLLocationAccuracyBest];
```

You need to know that regions are a shared system resource — the total number of regions available is limited. If you attempt to register a region and space is unavailable, the location manager sends the delegate the location Manager:monitoringDidFailForRegion:withError: message with the kCLErrorRegionMonitoringFailure error code.

While monitoring of a region begins immediately, only boundary crossings can generate an event — if the user is inside the region already, you won't get an event until he or she leaves it.

When the user does enter or leave a region, the location manager sends a delegate message:

```
- (void)locationManager:(CLLocationManager *)manager
        didEnterRegion:(CLRegion *)region {

 // you could say something clever here
}
- (void)locationManager:(CLLocationManager *)manager
        didExitRegion:(CLRegion *)region {

  [self leavingHomeNotify];
}
```

(You won't be using locationManager:didEnterRegion:. I show it here so you understand that you can use this service to notify someone that they have entered a region as well as left one.)

If your application is already running, these events go directly to the delegates of any current location manager objects. If your application is not running, the system launches it in the background so that it can respond.

Figure 18-6 shows what happens after I've left my house and left the region I defined in Listing 18-10. As you can see, I defined a circular region of 100 meters.

Figure 18-6:
Don't leave home without the essentials!

Your Application May Be Launched

This brings me to the last thing you really need to know about multitasking (really).

If your application processes events in the background, or if you are using the significant-change or region-monitoring services, if the application is terminated the system relaunches the application automatically when new location data becomes available.

If you've been wondering what all those options are for in the `application: didFinishLaunchingWithOptions:` method, here's where you find out. If your application was launched by the system for a specific reason, the `launchOptions` dictionary is going to contain data indicating the reason for the launch. The dictionary contains keys for a number of events, but I'm going to limit my discussion to the following two:

- **The application was launched by the user in response to the arrival of a local notification.** The key for this event is `UIApplicationLaunchOptionsLocalNotificationKey`.

- **The application tracks location updates in the background, was purged, and has now been relaunched.** The key for this event is `UIApplicationLaunchOptionsLocationKey`

As you see so far, in iPhoneTravel411 you really don't need to know all your launch options in order to process location or events or local notifications. But circumstances may arise in your application that make knowing such details a necessity. Listing 18-11 shows you how to check for a launch that comes about as a result of a local notification.

Listing 18-11: Launched Due to a Local Notification

```
UILocalNotification *locationNotification = [launchOptions
        objectForKey:UIApplicationLaunchOptionsLocation
        Key];
  if (locationNotification) {
    UILocalNotification *note = [[UILocalNotification
        alloc] init];
    note.alertBody= @"iPhoneTravel441 was launched to
        tell you don't forget your iPhone, tickets, and
        passport";
    [[UIApplication sharedApplication] presentLocalNotific
        ationNow:note];
    [note release];
  }
```

Figure 18-7 shows the result.

Figure 18-7:
Launched to
respond to
an event.

Requiring the Presence of Location Services in Order to Run

If your application relies on location services to function properly, you should include the UIRequiredDeviceCapabilities key in the application's Info.plist file. You use this key to specify the location services that must be present in order for your application to run. The App Store uses the information in this key to prevent users from downloading applications to devices that don't contain the listed features.

The value for the `UIRequiredDeviceCapabilities` is an array of strings indicating the features that your application requires. The following two strings are relevant to location services:

- ✔ `location-services`: Include this string if you require location services in general.
- ✔ `gps`: Include this string if your application requires the accuracy offered only by GPS hardware.

If your application makes use of location services but can run perfectly well without them, don't make such services a requirement by using the `UIRequiredDeviceCapabilities` key.

What's Next?

Although this point marks the end of your guided tour of iPhone software development, it should also be the start — if you haven't started already — of your own development work.

Developing for the iPhone is one of the most exciting opportunities I've come across in a long time. I'm hoping it ends up being as exciting — and perhaps less stressful — an opportunity for you.

Do keep in touch, though. Check out my Web site, www.nealgoldstein.com, on a regular basis. There you can find the completed Xcode projects for ReturnMeTo and iPhoneTravel411. You can also ask me questions, and if I find that more than a few people are confused about something, I'll post a clarification on my site as well. I'll also update things periodically, including insights from app developers about the best way to sell apps, new features from Apple, or policy changes on what app developers may and may not do in their apps. This last one is very important because Apple polices can be very fluid.

I'll also keep you up-to-date on where mobile apps are heading in general, how industry trends are shaping up, and what The Next Big Thing may look like.

Finally, keep having fun. I hope I have the opportunity to download one of your applications from the App Store someday.

Part V
The Part of Tens

The 5th Wave   By Rich Tennant

"Okay, the view's just up ahead. Everyone switch
to 'America the Beautiful' on your
iPhone playlist."

In this part . . .

I once had a boss who liked to hire smart-but-lazy people. He figured if he gave them a hard job, they'd find an easy way to do it.

In this part, I show you some ways to (first) avoid doing more work than you have to, and (second) avoid redoing things because you outsmarted yourself. I take you on a tour of Apple's sample applications and point out where to look if you want to "borrow" some code to implement some piece of functionality. (Don't worry — it's strictly legit to do that.) Then I show you where spending some up-front time doing app development the right way is definitely worth it.

Chapter 19

Top Ten Apple Sample Applications (With Code!)

· ·

"Good artists copy. Great artists steal."

– Pablo Picasso

One way to really find out how to do things on the iPhone is to look at (correct) sample code. Apple provides a lot of it. The only problem with learning from samples is that, even though it can show you how to do a specific thing (like flip a view), it doesn't give you the overall architectural understanding you need to create an application.

After going through this book, you know enough to take real advantage of all this sample code. By all means, take what you can from the samples and use it where you can to jump-start your own application development.

Here are the ten samples I like the best. You can find them at the iPhone Dev Center Web site at `http://developer.apple.com/iphone` by clicking the Sample Code link. You might have to scroll down to find the link.

They all come as Xcode projects, so you can check out the code, compile the project, and then load it into the simulator or even onto your development iPhone.

AppPrefs

If you want to get a good handle on preferences and how to set, access, and use them in your application, this is the place to get that grip. This app even has a table view with flip animation that explains the application on the flip side.

BubbleLevel

I can't help it. I love this application. It includes a lot of things any iPhone app developer would want to know how to do, including graphics and audio, and how to develop an application with a landscape view. It also allows you to calibrate the level with directions on the flip side of the level view. I actually keep this app on my iPhone. You never know when you might need a level to straighten a picture or build a deck.

WorldCities

While WorldCities demonstrates the basic use of the `MapKit` framework, including displaying a map view and setting its region — something you already know how to do — it adds some functionality you might want to include in your app.

The list of cities is stored in a `plist` file loaded at launch time. It processes the `plist` file to create a `WorldCity` class, which consists of a name, a latitude, and a longitude. The list of `WorldCity` entries is then used in a table view to create a selectable list; when you select a city, the map view animates to a region with the coordinates of the selected world city in the center of the view. It also shows you how to add a segmented control in the toolbar of the main view that enables the user to choose from the available map types: Standard, Satellite, and Hybrid. I do similar things in my own apps, and this is a handy way to get started.

QuartzDemo

Quartz2D is the foundation of all drawing in the iPhone OS — and is something I don't cover in this book. This framework provides facilities for drawing lines, polygons, curves, images, gradients, PDF document generation, and many other graphical facilities. QuartzDemo, as you might guess, demonstrates many of the Quartz2D APIs and is a good place to start finding out about how to draw on the iPhone.

Reachability

If you use an Internet connection in your app, you need to be able to determine whether a connection is available so you can then inform the user that a function is not available when you find there is no Internet connection. You can use this code to check whether a particular site is available. This app also shows you, if you uncomment out a line of code, how to run in asynchronous mode and notify the application of a change in device status.

iPhoneCoreDataRecipes

This sample shows you how to do a lot of things; among them, using view controllers, table views, and Core Data in an iPhone application. It uses the view controller to manage information, and table views to display and edit data. What's more important, though, is its use of Core Data. Core Data is the way you should think about saving objects and then retrieving them from a persistent store (such as a file on disk), and the sample shows you how to implement a Core Data persistent store.

UICatalog

This sample illustrates all the elements you'd want to use in your app's user interface. (You know — all those buttons, icons, and bangles that the user sees onscreen.) This one application shows you how to implement all of them.

URLCache

URLCache demonstrates how to download a resource, store it in the data directory, and use the local copy. All those tasks are very useful for anything you want your app to do on the Web. It also includes a framework for asynchronous processing. If you need to download a lot o' data when the application starts, this sample shows you how to tell your app to start a download and then go off and do other things until the download completes. It also includes an activity-indicator view that tells the user your application is actually at work, and not on a lunch break.

XML

There are actually two XML samples I can recommend. The first one, SeismicXML, shows you how to work with XML documents. When you launch it, it gets and parses an RSS feed. This particular feed is from the U.S. Geological Survey that provides data on recent earthquakes around the world. (I live in Northern California, so you can bet I keep this one on my iPhone as well.)

The second sample, XMLPerformance, parses XML by using the two APIs provided in the SDK. This sample allows the user to choose between the two APIs, tracks the statistics of each parse, and stores that data in an SQLite database. There's a good discussion of performance in the ReadMe file. In addition, the app's RSS feed uses the "Top 300" songs from iTunes, so you can keep up-to-date on those tunes.

Tables

Numerous samples can show you how to implement the functionality inherent in the table view. I just list them here, and you can examine them at your leisure:

- ✔ **Accessory** implements a checkmark button in a custom accessory view.

- ✔ **AdvancedTableViewCells** shows you how to create three different cells (and ways) that all display content in the same form as the AppStore application. This includes subviews (image views, labels, and so on), a single view to draw most of the content, as well as separate views for the remainder.

- ✔ **DateCell** shows you how to format date objects in a table-view cell and then use a date picker to edit the dates.

- ✔ **DrillDownSave** saves the current location in a hierarchy and then restores the current location when the user relaunches the application. (Hmm, I wonder where I've seen that before?)

- ✔ **HeaderFooter** shows customized header and footer views.

- ✔ **LazyTableImages** shows you how to use a multi-stage approach to load and display data in a `UITableView`. It starts by loading the relevant text from an RSS feed so the table can load as quickly as possible, and then it downloads the images for each row asynchronously to keep the UI responsive.

- ✔ **SimpleDrillDown** is a simple drill-down application using a `UITableView`.

- ✔ **TableSearch** implements searching by using `UISearchBar` and `UISearchDisplayController` and then filtering content. This is similar to what the Mail application does when you start to type an e-mail address in the To field when you are composing an e-mail.

- ✔ **TableViewSuite** shows a lot of table views.

- ✔ **TheElements** is a very robust application — and is structured as a Model-View-Controller application to boot. It allows you to sort data and present it in multiple formats. It uses a tab bar, displays in plain and grouped table views, uses navigation controllers to navigate deeper into a data structure, creates custom table-view cells with multiple subviews, accesses a Web site by using Safari, reacts to taps in a view, flips view content from front to back, and reflects a view. Makes me tired just to think about it.

- ✔ **TouchCells** implements controls in a table view.

Chapter 20

Ten Ways to Be a Happy Developer

In This Chapter

▶ Finding out how not to paint yourself into a corner

▶ Avoiding "There's no way to get there from here."

There are lots of things you know you're supposed to do, but you don't because you think they'll never catch up with you. (After all, not flossing won't cause you problems until your teeth fall out years from now, right?)

But in iPhone application development, those things catch up with you early and often, so I want to tell you about what I've learned to pay attention to from the very start in app development, as well as a few tips and tricks that lead to happy and healthy users.

It's Never Early Enough to Start Speaking a Foreign Language

With the world growing even flatter, and the iPhone available in more than 80 countries, the potential market for your app is considerably larger than just people who speak English. Localizing an application isn't difficult, just tedious. Some of it you can get away with doing late in the project, but when it comes to the strings you use in your application, you better build them right — and build them in from the start. The painless way: Use the `NSLocalizedString` macro (refer to Chapter 14) from the very start, and you'll still be a happy camper at the end.

Remember Memory

The iPhone OS does not store "changeable" memory (such as object data) on the disk to free up space and then read it back in later when needed. It also doesn't have garbage collection — which means there's a real potential for memory leaks unless you tidy up after your app. Review and follow the memory rules in Chapter 6 — in particular, these:

- ✔ Memory management is really creating *pairs* of messages. Balance every `alloc`, `new`, and `retain` with a `release`.
- ✔ When you assign an instance variable by using an accessor with a property attribute of `retain`, you now own the object. When you're done with it, release it in a `dealloc` method.

Constantly Use Constants

In the iPhoneTravel411 application, I put all my constants in one file. When I was developing the MobileTravel411 projects, I did the same. The why of it is simple: As I changed things during the development process, having *one* place to find my constants made life much easier.

Don't Fall Off the Cutting Edge

The iPhone is cutting-edge enough that there are still plenty of opportunities to expand its capabilities — and many of them are (relatively) easy to implement. You're also working with a very mature framework. So if you think something you want your app to do is going to be really difficult, check the framework; somewhere in there you may find an easy way to do what you have in mind. If there isn't a ready-made fix, consider the iPhone's limited resources — and at least question whether that nifty task you had in mind is something your app should be doing at all. Then again, if you really *need* to track orbital debris with an iPhone app, go for it — someone needs to lead the way. Why shouldn't it be you?

Start by Initializing the Right Way

A lot of my really messy code that I found myself redoing ended up messy because I didn't think through initialization. (For example, adding on initialization-like methods after objects are already initialized is a little late in the game, and so on.) Reread and heed Chapter 16; the initialization process is important in implementing reusable view controllers and models.

Keep the Order Straight

One of the things that can really foul up your day as a developer is the order in which objects are called. If you expect an object to be there (and it isn't) or to have been initialized (and it wasn't), you may be in the wrong method. Copy Table 20-1 and paste it into a file — and/or print it out and tack it up where you can easily find it.

Table 20-1	The Natural Order of Things
Object	*Method*
View controller	`awakeFromNib`
Application delegate	`application:didFinishLaunchingWithOptions:`
View controller	`viewDidLoad`
View controller	`viewWillAppear:`
View controller	`viewWillDisappear:`
Delegate	`applicationWillTerminate:`

What trips up many developers is that the `awakeFromNib` message for the initial view controller (the one you see when the application starts) is sent *before* the `application:didFinishLaunchingWithOptions::` message. If you have a problem with that, do what you need to do in `ViewDidLoad`.

You also need to pay attention to the methods called as you leave the foreground, enter the background, and then return to the foreground again. I cover those extensively in Chapter 2.

Avoid Mistakes in Error Handling

A lot of opportunities for errors are out there; use common sense in figuring out which ones you should spend work time on. For example, don't panic over handling a missing directory in your code. On the iPhone, it's supposed to be there; if it's not, look for a bug in your program. If it's *really* not there, then the user has big problems, and you probably won't be able to do anything to avert the oncoming hassle.

There are, however, some potential pitfalls you do have to pay attention to, such as these two big ones:

  ✔ Your app goes out to load something off the Internet, and (for a variety of reasons) the item isn't there or the app can't get to it. You especially need to pay attention to Internet availability and what you're going to do when the Internet isn't available.

✔ An object can't initialize itself (for a similar range of perverse reasons).

When, not if, those things happen, your code and your user interface must be able to deal with the error.

Remember the User

I've been singing this song since Chapter 1, and I'm still singing it now: Keep your app simple and easy to use. Don't build long pages that take lots of scrolling to get through, and don't create really deep hierarchies. Focus on what the user wants to accomplish, and be mindful of the device limitations, especially battery life. And don't forget international roaming charges.

In other words, try to follow the Apple's iPhone Human Interface Guidelines, found with all the other documentation in the iPhone Dev Center Web site at `http://developer.apple.com/iphone` under the iPhone Reference Library section — Required Reading. Don't even *think* about bending those rules until you really, *really* understand them.

Keep in Mind that the Software Isn't Finished Until the Last User Is Dead

If there's one thing I can guarantee about app development, it's that Nobody Gets It Right the First Time. The design for MobileTravel411 evolved over time, as I learned the capabilities and intricacies of the platform and the impact of my design changes. Object orientation makes extending your application (not to mention fixing bugs) easier, so pay attention to the principles.

Keep It Fun

When I started programming the iPhone, it was the most fun I'd had in years. Keep things in perspective: Except for a few tedious tasks (such as provisioning and getting your application into the Apple Store), lo, I prophesy, developing iPhone apps will be fun for you, too. So don't take it *too* seriously.

Especially remember the *fun* part at 4 a.m., when you've spent the last five hours looking for a bug.

Index

• E •

• X •

• Z •

Notes